Conference Board of the Mathematical Sciences
REGIONAL CONFERENCE SERIES IN MATHEMATICS

Number 11

Irving Glicksberg

Recent results on function algebras

Supported by the National Science Foundation
Published by the American Mathematical Society

Other Monographs in this Series

No. 1. Irving Kaplansky: *Algebraic and analytic aspects of operator algebras*

2. Gilbert Baumslag: *Lecture notes on nilpotent groups*

3. Lawrence Markus: *Lectures in differentiable dynamics*

4. H. S. M. Coxeter: *Twisted honeycombs*

5. George W. Whitehead: *Recent advances in homotopy theory*

6. Walter Rudin: *Lectures on the edge-of-the-wedge theorem*

7. Yozo Matsushima: *Holomorphic vector fields on compact Kähler manifolds*

8. Peter Hilton: *Lectures in homological algebra*

9. I. N. Herstein: *Notes from a ring theory conference*

10. Branko Grünbaum: *Arrangements and spreads*

Conference Board of the Mathematical Sciences
REGIONAL CONFERENCE SERIES IN MATHEMATICS

supported by the

National Science Foundation

Number 11

RECENT RESULTS ON FUNCTION ALGEBRAS

by

IRVING GLICKSBERG

Published for the

Conference Board of the Mathematical Sciences

by the

American Mathematical Society

Providence, Rhode Island

Expository Lectures
from the CBMS Regional Conference
held at Fayetteville, Arkansas
June 1971

International Standard Book Number 0-8218-1658-6
Library of Congress Catalog Card Number 75-38927
AMS 1969 Subject Classifications: 3070, 4655, 4687

Printed in the United States of America

FOREWORD

What follows are notes covering lectures given at a Regional Conference at Fayetteville, Arkansas in June 1971, sponsored by CBMS and the NSF. My aim in these lectures was to introduce some of the more recent developments on function algebras, presuming no detailed knowledge on the listener's part. Of course then it was necessary to develop some of the foundations, and some older results are interspersed among the newer ones. Some areas are also slighted considerably: for example, we shall have nothing to say about analytic structure in the spectrum; and because of the technical aspect, we will really cover only one result in rational approximation in the plane (an area where considerable recent progress has been made via a combination of the classical and abstract techniques), and will wholly ignore higher dimensions.

Here, briefly, is what we attempt to cover: (1) some basic notions, results and examples of uniform algebras, (2) interpolation, (3) orthogonal measures, (4) rational approximation, and (5) recent characterizations of $C(X)$.

I am greatly indebted to various friends for kindly providing preprints and the latest and simplest versions of proofs, particularly A. Bernard, B. Cole, A. M. Davie, T. W. Gamelin, J. Garnett, S. J. Sidney, and N. T. Varopoulos. Finally, thanks are due the Department of Mathematics of the University of Arkansas, and especially Professor Allan Cochran, without whose untiring efforts the conference could not have taken place.

1. UNIFORM ALGEBRAS: SOME BASIC NOTIONS, RESULTS, AND EXAMPLES

We shall be mainly concerned with uniform (or sup normed) algebras: closed subalgebras A of $C(X)$, X compact Hausdorff, which contain the constants. Frequently we will assume A separates the points of X, and almost always A will not be closed under conjugation.

The simplest and best example is the disc algebra $A = A(D)$ consisting of all continuous f on the closed disc D in $\mathbf{C}$ analytic on its interior D°. From the maximum principle we know A and $A \mid T$, $T = \partial D$, are isometrically isomorphic objects, and we have our choice of investigating A on D, or T, or any set between, so that there are many candidates for X. And of course a particular X may enjoy special properties, as for example in

WERMER'S MAXIMALITY THEOREM 1.1. *$A = A(D) \mid T$ is a maximal (proper) closed subalgebra of $C(T)$.*

Trivially this is false if T is replaced by any larger subset of D. Before proceeding, we will note a simple recent proof (due to Lumer), since it provides a good simple example of the use of functional analytic tools. We only have to know that (i) the open ball of unit radius about the identity in a Banach algebra consists of invertibles (for the proof, write the obvious power series for $(1-x)^{-1}$), and (ii) $\int_0^{2\pi} e^{in\theta} f(e^{i\theta})d\theta = 0$ for $n \geqslant 1$ implies $f \in A$ (for $f \in C(T)$).

Suppose $A \subset B \subset C(T)$, with B closed. If $e^{-i\theta} \in B$ then of course $B = C(T)$ by Stone-Weierstrass. Thus we can assume $\|e^{-i\theta} - B\| = \inf \|e^{-i\theta} - b\| > 0$, or $\|1 - e^{i\theta} b\| > 0$ for all $b \in B$. And even if we had $\|1 - e^{i\theta} b\| < 1$ for some b then $e^{i\theta} b \in B^{-1}$ (the set of invertibles in B), so $(e^{-i\theta} b^{-1}) \cdot b = e^{-i\theta} \in B$ again. Thus $\|1 - e^{i\theta} b\| \geqslant 1$ for all b, or $\|1 - e^{i\theta} B\| = 1$.

Now by the Hahn-Banach and Riesz representation theorems we have a complex measure μ of norm 1 on T which is 1 at 1 and orthogonal to $e^{i\theta}B$ (via the fact, say, that the dual $(C(T)/e^{i\theta}B)^*$ is a subspace of $M(T) = C(T)^*$). But $\mu(1) = 1 = \|\mu\|$ implies μ is a probability measure and since $e^{in\theta} \in e^{i\theta}B$, $n \geqslant 1$, $\int e^{in\theta} \mu(d\theta) = 0$, $n \geqslant 1$, hence also for $n \leqslant -1$, and $\mu(d\theta) = d\theta/2\pi$. But now

$$\mu(e^{in\theta} b) = \mu(e^{i\theta} e^{i(n-1)\theta} b) = 0, \quad n \geqslant 1,$$

implies $b \in A$, as we know from (ii).

In our example of the disc algebra, D is the spectrum $\Delta = \Delta_A$ (or maximal ideal space, or space of non-trivial, i. e. non-zero, multiplicative linear functionals (m. l. f.) on A), and T is the smallest closed *boundary* for A: a set on which each element assumes its maximum modulus. (Of course point evaluations provide the multiplicative linear functionals.) We can always obtain an isometrically isomorphic object restricting A to any boundary X. In general if $1 \in A$, a closed subalgebra of $C(X)$, then any m. l. f. ϕ on A has norm $\leqslant 1$; if B is the open unit ball in A, then, since $1 + B \subset A^{-1}$, $0 \notin \phi(1 + B) = 1 + \phi(B)$, and, since $\phi(B)$ cannot contain -1, it clearly cannot contain any point of T, whence $\phi(B) \subset D$. By Hahn-Banach, ϕ has an extension to $C(X)$ of norm $\leqslant 1$, which is then given by a measure λ on X, with $\lambda(1) = 1 = \|\lambda\|$. So we have a probability measure λ on X with

$$\phi(a) = \int a d\lambda, \quad a \in A. \tag{1.1}$$

Such a measure is called a representing measure for ϕ. (A complex λ for which (1.1) holds will be referred to as a *complex* representing measure; finally by a multiplicative measure one means any λ satisfying (1.1) for some ϕ.)

Evidently λ might just be a point mass, but only if ϕ is evaluation at a point of X, and otherwise it will not be that trivial. It may or may not be unique: for $A = A(D)|T$,

$$\phi(a) = a(0) = \int a(e^{i\theta})\, d\theta/2\pi$$

and our measure is unique as we have noted; but, for the analogous bidisc algebra $A(D^2)$ (the continuous f analytic on $(D^2)^\circ$) restricted to T^2,

$$a(0, 0) = \iint a(e^{i\theta}, e^{i\phi})\, d\theta d\phi/4\pi^2 = \int a(e^{i\theta}, e^{i\theta})\, d\theta/2\pi.$$

We shall let M_ϕ denote the full set of representing measures for a m. l. f. ϕ (assuming one fixed X); since we have a probability measure satisfying (1.1) we in fact have

$$\operatorname{Re} \phi(a) = \int \operatorname{Re} a\, d\lambda, \quad a \in A, \tag{1.2}$$

which also implies (1.1) since $\operatorname{Im} a = \operatorname{Re}(-ia)$, and which further shows $\operatorname{Re} a \longrightarrow \operatorname{Re} \phi(a)$ is a well-defined non-negative real linear functional on $\operatorname{Re} A$. So, as a simple consequence of the order-preserving form of Hahn-Banach, one has for $u \in C_R(X)$

$$\sup\{\operatorname{Re} \phi(a) : \operatorname{Re} a \leqslant u,\ a \in A\} = \inf\{\lambda(u) : \lambda \in M_\phi\}. \tag{1.3}$$

(Indeed, if c denotes the left side, then $\operatorname{Re} a \leqslant u \Rightarrow \lambda(\operatorname{Re} a) = \operatorname{Re} \phi(a) \leqslant \lambda(u)$ shows it is no greater than the right side, while $ru + \operatorname{Re} a \longrightarrow rc + \operatorname{Re} \phi(a)$ is easily seen to be a non-negative extension of $\operatorname{Re} a \longrightarrow \operatorname{Re} \phi(a)$ to $Ru + \operatorname{Re} A$. A Hahn-Banach extension of that to $C_R(X)$ yields a $\lambda \geqslant 0$ with $\lambda(1) = 1$ of course and $\lambda(u) = c$, so equality obtains in (1.3).) We will have use of (1.3) later.

Our proof of Wermer's Maximality Theorem 1.1 is the first of many uses of the functional approach; but sometimes we shall exploit the algebraic aspects of our setting, as in the following result, which we include because it is so easy and combines with 1.1 to give a well-known result.

LEMMA 1.2. *If $f \in A$ vanishes on the relatively open subset U of a boundary $Y \subset X$, then f vanishes outside $\widehat{Y \setminus U} = \{x \in X : |a(x)| \leqslant \sup |a(Y \setminus U)|,\ a \in A\}$ (thus if* *this last set is proper,* f *vanishes on an open set in* X).

Our proof is due to Wermer. Fix $x \in X$ with $f(x) \neq 0$. Then, for $g \in A$,

$$|fg(x)| \leqslant \sup |fg(Y)| = \sup |fg(Y \backslash U)| \leqslant \sup |f(Y \backslash U)| \sup |g(Y \backslash U)|,$$

so $|g(x)| \leqslant c \sup |g(Y \backslash U)|$, where $c = |f(x)|^{-1} \sup |f(Y \backslash U)|$. Replacing g by g^n we have $|g(x)|^n \leqslant c(\sup |g(Y \backslash U)|)^n$ so that taking nth roots and letting $n \longrightarrow \infty$,

$$|g(x)| \leqslant \sup |g(Y \backslash U)|$$

and $x \in \widehat{Y \backslash U}$, as desired.

COROLLARY 1.3 (RADÓ). *If* $f \in C(D)$ *is analytic except where it is zero,* f *is analytic on* D°.

Let B be the closed subalgebra of $C(D)$ generated by f and $A = A(D)$ and take D as our X. Every $g \in B$ is analytic on $D^\circ \backslash f^{-1}(0)$, and on $f^{-1}(0)^\circ$, so that by the maximum principle $T \cup \partial f^{-1}(0) = Y$ is a boundary for B, and f vanishes on the relatively open subset $U = \partial f^{-1}(0) \backslash T$ of Y. By the lemma, with B our algebra, f vanishes outside $\widehat{Y \backslash U} = \hat{T}$ so that every point in $D \backslash f^{-1}(0)$ lies in $\hat{T}$. But then so does $\partial f^{-1}(0)$ since $\hat{T}$ is closed, so that $Y \subset \hat{T}$ and T is a boundary for B.

Now $B | T = C(T)$ or $B | T = A | T$ by 1.1. The first cannot obtain since evaluation at any point z_0 of D° yields a m. l. f. on B, hence a representing measure λ on T, so that $(z - z_0)\lambda(dz)$ is a non-zero measure on T orthogonal to B. Consequently, A and B coincide on T, and if $a = f$ on T, $a \in A$, then of course $a - f = 0$ on D since T is a boundary for B.

Radó's theorem gives one answer, in a very special case, to the general question: what conditions will insure that an f in $C(X)$ (or $C(\Delta)$) lies in A? More generally, and quite importantly, given two algebras, when do they coincide, and how can we determine this? For example, if $K \subset \mathbf{C}$ is compact, and $A(K)$ consists of the f in $C(K)$ analytic on K°, while $R(K)$ consists of just the uniform closure of those f analytic near K (so $R(K) \subset A(K)$), one of the central problems of rational approximation is to give satisfactory necessary and sufficient conditions for their equality.* We shall look at this in more detail later, turning next to an old result (giving a necessary and sufficient condition for a single f to belong to A) which provides another simple example of the functional approach.

THEOREM 1.4. BISHOP'S GENERALIZED STONE-WEIERSTRASS THEOREM. *Call a set* $K \subset X$ *a set of antisymmetry (a. s.) for* A *if* $A | K$ *contains no non-constant real functions. Then the set* $\mathcal{K}$ *of maximal sets of antisymmetry forms a closed pairwise disjoint covering of* X *for which*

(1.4) $f \in C(X)$, *and* $f | K \in (A | K)^-$ *for all* $K \in \mathcal{K}$ *imply* $f \in A$. *(The bar denotes closure in* $C(K)$.*)*

---

* By the well-known classical result of Runge, $R(K)$ is the uniform closure on K of the rational functions with poles off K, whence the notation and terminology.

Trivially a point is an a. s., as is the union of any two a. s. which meet. Thus the union of all a. s. containing a given point is an a. s. and necessarily maximal, and closed since its closure is also an a. s. So our first assertion is clear. For the second, to show $f \in A$ it is enough to show f is orthogonal to μ in $A^{\perp}$, (*always*) the set of complex measures on X orthogonal to A. And by Krein-Milman that means it suffices to show f orthogonal to all $\mu \in (\text{ball } A^{\perp})^e$, the set of extreme points of the unit ball of $A^{\perp}$. But such a μ has its support K an a. s.: for if $g \in A$ is real on K and $0 < g < 1$ there, then

$$\mu = \|g\mu\| \frac{g\mu}{\|g\mu\|} + \|(1-g)\mu\| \frac{(1-g)\mu}{\|(1-g)\mu\|}$$

expresses μ as a convex combination of two elements of ball A since

$$\|g\mu\| + \|(1-g)\mu\| = \int gd\,|\mu| + \int (1-g)d\,|\mu| = \|\mu\| = 1;$$

thus g must be constant mod $|\mu|$, and so constant on K. Of course there is no loss of generality in assuming $0 < g < 1$ on K since $1 \in A$, so K is antisymmetric as asserted. Trivially (1.4) implies $\mu \perp f$ and we are done.

Actually $A\,|\,K$ is closed, but we shall have no use for even the weaker result we have given. We shall have use, however, for the basic result from which closure follows.

DEFINITION. *A closed $F \subset X$ is called a peak set (for A) if some $f \in A$ has $f(F) = 1$, $|f| < 1$ off F, and then f is said to peak on F.*

A "generalized" peak set is an intersection of peak sets, thus a set with the property that between it and any of its neighborhoods, we can fit a genuine peak set. There is of course no distinction in case X is metric, and at times we may want to stick to the metric case to avoid the slight complication in passing from peak sets to generalized peak sets. We want two fundamental results on peak sets due to Bishop.

BISHOP'S PEAK SET LEMMA 1.5. *If F is a peak set for A and $F_0 \subset F$ is a peak set for $A\,|\,F$, then F_0 is also a peak set for A (i. e., if $f(F_0) = 1$ while $|f| < 1$ on $F \backslash F_0$, $f \in A$, there is a $h \in A$ with $h(F_0) = 1$, $|h| < 1$ on $X \backslash F_0$).*

Suppose $f(F_0) = 1$ and $|f| < 1$ on $F \backslash F_0$, and $g(F) = 1$ while $|g| < 1$ on $X \backslash F$, $f, g \in A$. We can assume $|f| < 3/2$ on X, replacing f by $(f + u)/(1 + u)$ for $u \in R$ large, if necessary. Let

$$K_n = \{x \in X : 1 + 2^{-n-1} \leqslant |f(x)| \leqslant 1 + 2^{-n}\},$$

$n = 1, \cdots$. Then $K_n \cap F = \emptyset$ so that there is an integer k_n with $|g^{k_n}| < 2^{-2n}$ on K_n. Setting $h = \Sigma\, 2^{-n} f g^{k_n} \in A$, we have $h(F_0) = 1$ and $|h(x)| \leqslant |f(x)| < 1$ for $x \in F \backslash F_0$; indeed clearly $|h(x)| < 1$ for any $x \notin F$ with $|f(x)| \leqslant 1$. On the other hand if $|f(x)| > 1$, so that $x \in K_m$ for some $m \geqslant 1$, then

$$|h(x)| \leqslant |f(x)| \left(\sum_{n \neq m} 2^{-n} + 2^{-m}\, 2^{-2m} \right)$$
$$\leqslant (1 + 2^{-m})\ (1 - 2^{-m} + 2^{-3m}) =$$

$$= 1 - 2^{-2m} + 2^{-3m} + 2^{-4m} = 1 - \frac{2^{2m} - 2^m - 1}{2^{4m}} < 1,$$

whence $|h(x)| < 1$ for all $x \notin F_0$.

Note: More generally this shows any element $f|F$ of $A|F$ has an extension h in A with $|h(x)| < \sup|f(F)|, x \notin F$.

For any boundary X for A we have a natural continuous imbedding of X into Δ (Δ carries the least fine topology rendering the functions $\phi \longrightarrow \phi(a)$ continuous). If A separates the points of X we thus have X imbedded homeomorphically in Δ; considering boundaries as subsets of Δ then they are comparable, and 1.5 shows there is a smallest (closed) boundary.

COROLLARY 1.6. *Let P denote the set of (generalized) peak points in Δ. Then P^- is the smallest closed boundary for A in Δ (the "Šilov boundary").*

As we will see in a moment, a peak point in X remains a peak point when X is imbedded in Δ and "peak point" has a meaning quite independent of which boundary we may take as our X. The proof of 1.6 is simple: each $f \in A$ assumes a value of maximum modulus (say 1) on a peak set (where $(1 + f)/2$ peaks), and by Zorn (and 1.5!) that set contains an element of P, so P^- is a boundary. Evidently $p \in P$ must be in each closed boundary, and we are done.

In order to see that a (generalized) peak point x_0 in X remains one in Δ, note that x_0 is an intersection $\bigcap f^{-1}(1) = \bigcap \{x \in X: f(x) = 1\}$ over a family of f in A of norm 1 for which $|f(y)| < 1$ unless $f(y) = 1$. But the corresponding intersection $\bigcap \{x \in \Delta: f(x) = 1\}$ is precisely $\{x_0\}$; for if ϕ is in that intersection and λ is a representing measure for ϕ carried by X then $\phi(f) = \lambda(f) = 1$ shows λ is carried by $\{x_0\}$ itself.

Our next result is Bishop's $\frac{1}{4} - \frac{3}{4}$ criterion.

THEOREM 1.7. *Suppose A is a linear subspace of $C(X), x \in X,$ and for each neighborhood U of x there is an f in A of norm 1 with $|f(x)| > \frac{3}{4}, |f(U')| < \frac{1}{4}$. Then x is a (generalized) peak point for A.* (U' denotes the complement of U.)

Alternatively we can suppose there is an f in A of norm $4/3$ with $f(x) = 1$, $|f(U')| < 1/3$. We have to show that for any neighborhood U_1 of x there is a peak set containing x and lying within U_1. Let f_1 be an element of norm $\leqslant 4/3, f_1(x) = 1$, $|f(U_1')| < 1/3$. Having chosen $U_1, \cdots, U_{n-1}, f_1, \cdots, f_{n-1}$, we set

$$U_n = \left\{x \in U_{n-1} : |f_j(x)| < 1 + \frac{1}{3}\, 2^{-n},\ j \leqslant n-1\right\}$$

and choose a corresponding f_n as in our hypothesis. We set $f = \Sigma_1^\infty 2^{-n} f_n \in A$.

If $y \notin U_1$ then $|f_n(y)| < 1/3$ for all n and $|f(y)| < 1/3$. If $y \in U_n \setminus U_{n+1}$ then $|f_j(y)| < 1/3$ for $j \geqslant n+1$ and $|f_j(y)| < 1 + 2^{-n}/3,\ j < n,$ so

$$|f(y)| \leqslant \sum_{1}^{n-1} 2^{-j}(1 + \frac{1}{3}\, 2^{-n}) + \frac{4}{3}\, 2^{-n} + \frac{1}{3} \sum_{n+1}^{\infty} 2^{-j}$$
$$= (1 - 2^{-n+1})\,(1 + \frac{1}{3}\, 2^{-n}) + \frac{4}{3}\, 2^{-n} + \frac{1}{3}\, 2^{-n}$$
$$< 1 - 2 \cdot 2^{-n} + \frac{1}{3}\, 2^{-n} + \frac{5}{3}\, 2^{-n} = 1.$$

Finally $y \in \bigcap U_n$ implies $|f_j(y)| \leqslant 1$ for all j, so $|f(y)| \leqslant 1$. Thus $x \in f^{-1}(1) \subset U_1$, $\|f\| = 1$, and $(1 + f)/2$ peaks on $f^{-1}(1)$.

COROLLARY 1.8. *For an algebra $A \subset C(X)$ containing* **C** *the following are equivalent*:

(a) *x is a (generalized) peak point for A in X*;

(b) *δ_x, the unit mass at x, is the only representing measure for x*;

(c) *any λ representing x has $\lambda\{x\} > 0$*;

(d) *for any neighborhood U of x there is an f in A with* $\|f\| = 1$, $|f(x)| > 3/4$, $|f(U')| < 1/4$.

We know (d) ⇒ (a); we will show (a) ⇒ (b) ⇒ (c) ⇒ (b) ⇒ (d).

If F is a peak set and $\mu \perp A$ then $\mu_F \perp A$ (if $h \in A$ peaks there, then $h^n\mu \longrightarrow \mu_F$ in norm by dominated convergence, and $h^n\mu \perp A$). Indeed the same conclusion holds for a generalized peak set by regularity. So $\delta_x - \lambda \perp A$ for any λ representing x implies $(\delta_x - \lambda)_{\{x\}} = (1 - \lambda\{x\})\,\delta_x \perp A$ or $\lambda(x) = 1$ if x is a (generalized) peak point, whence $\lambda = \delta_x$ and (a) ⇒ (b). Clearly (b) ⇒ (c), and if $\mu \neq \delta_x$ represents x then so does $\lambda = (\mu - \mu\{x\}\delta_x)/(1 - \mu\{x\})$. So if (b) fails (c) does, and (c) ⇒ (b). We know (d) ⇒ (a), so we need only show (b) ⇒ (d). As we know from (1.3), for $u \in C_R(X)$ and M_x the representing measures for evaluation at x,

$$\sup\{\operatorname{Re} a(x) : \operatorname{Re} a \leqslant u,\ a \in A\} = \inf\{\lambda(u) : \lambda \in M_x\},$$

so that if (b) holds and $0 \leqslant u \leqslant 1 = u(x)$ and $u \equiv 0$ off U, then there is an a in A with $\operatorname{Re} a(x) > 1 - \epsilon$, $\operatorname{Re} a \leqslant u$. So $b = e^{2a}$ has $\|b\| \leqslant e^2$, $|b(x)| > e^{2-2\epsilon}$ and $|b(U')| \leqslant 1$, and $e^{-2}b$ provides our f in (d).

We need (a) ⇒ (c) in order to characterize peak sets in terms of orthogonal measures.

THEOREM 1.9. *Suppose A is a closed separating subalgebra of $C(X)$. Then a closed set $F \subset X$ is a generalized peak set if and only if $\mu \in A^\perp$ implies $\mu_F \in A^\perp$ (where $A^\perp$ denotes the measures on X orthogonal to A).*

We have already seen "only if", so suppose $\mu \in A^\perp$ implies $\mu_F \in A^\perp$. Let Y denote X with F identified to a point, $\rho : X \longrightarrow Y$ the corresponding map. We set $B = \{f \in C(Y) : f \circ \rho \in A\}$ so $B \circ \rho$ consists of the elements of A constant on K. Finally let $1^\perp(F)$ denote the measures on $F \perp 1$.

Clearly

$$(B \circ \rho)^\perp \supset A^\perp + 1^\perp(F) = A^\perp_{F'} + 1^\perp(F)$$

(where $A_{F'}^{\perp} = \{\mu_{F'} : \mu \in A^{\perp}\}$), and the right side is evidently w^* dense in the left, since any $f \in C(X)$ orthogonal to it lies in A and must be constant on F. So we have equality if the right side is w^* closed, and thus if its unit ball is w^* closed by the Kreĭn-Smulian Theorem. But if $\mu_\delta \in A_{F'}^{\perp}$, $\nu_\delta \in 1^{\perp}(F)$, $\|\mu_\delta + \nu_\delta\| \leqslant 1$ (so $\|\mu_\delta\| + \|\nu_\delta\| \leqslant 1$) and $\mu_\delta + \nu_\delta \longrightarrow \lambda$ w^*, then passing to a cofinal subnet we have $\mu_\delta \longrightarrow \mu \in A^{\perp}$, $\nu_\delta \longrightarrow \nu \in 1^{\perp}(F)$ (since these two spaces are w^* closed), so $\lambda \in A^{\perp} + 1^{\perp}(F)$ and the ball is w^* closed.

Hence we have $(B \circ \rho)^{\perp} = A_{F'}^{\perp} + 1^{\perp}(F)$. Of course this shows B is non-trivial (if $F \neq X$) and indeed big enough to separate Y: for if $\delta_{\rho(x)} - \delta_{\rho(y)} \perp B$ that means precisely that $\delta_x - \delta_y \perp B \circ \rho$, so

$$\delta_x - \delta_y = \mu + \nu, \qquad \mu \in A_{F'}^{\perp}, \nu \in 1^{\perp}(F),$$

and therefore $(\delta_x - \delta_y)(F) = 0$, whence either $x, y \in F$ and $\rho(x) = \rho(y)$, or $x, y \in F'$ and $\delta_x - \delta_y = \mu \in A^{\perp}$, so $x = y$ since A separates X.

Now it is enough to see $y_0 = \rho(F)$ is a (generalized) peak point for B (since then F is clearly a (generalized) peak set for $B \circ \rho$, hence for A). But if y_0 is not one, then by 1.8 we have a representing measure λ for y_0 with $\lambda\{y_0\} = 0$, and this λ is the image of a measure λ' on X obtained by setting $\lambda'(E) = \lambda(\rho(E))$. Of course $\lambda' - \delta_x \perp B \circ \rho$ for any x in F, so

$$\lambda' - \delta_x = \mu + \nu, \quad \mu \in A_F^{\perp}, \quad \nu \in 1^{\perp}(F),$$

and since $\lambda'(F) = \lambda\{y_0\} = 0$, $(\lambda' - \delta_x)(F) = -1$ while $\mu(F) + \nu(F) = 0 + 0$, we have our contradiction.

2. INTERPOLATION

If A is once more the disc algebra and $F \subset T$ a compact set of linear measure zero, Rudin and Carleson long ago proved $A|F = C(F)$, i.e., that F was a set of interpolation for A. Such sets, and indeed non-trivial ones, occur in the general setting. They are characterized quite easily in terms of $A^\perp$, the measures on X orthogonal to A, by an old result we will recall which evolved from an argument due to E. Nelson. (That result combines immediately with the F. and M. Riesz theorem asserting that a measure on T orthogonal to $A(D)$ is absolutely continuous, to yield the theorem of Rudin and Carleson.) Then we shall turn to recent results of Varopoulos and Dufresnoy on interpolation.

For F closed let $(A|F)^\perp$ denote the measures on F orthogonal to A.

THEOREM 2.1. *$A|F$ is closed in $C(F)$ iff there is a constant c for which*

(2.1) $$\|\mu - (A|F)^\perp\| \leqslant c\|\mu - A^\perp\| \quad \textit{for all } \mu \in M(F).$$

This is a simple consequence of the fact [D-S, p. 488] that a bounded linear map T from one Banach space to another has closed range iff T^* has. Indeed $T: f \longrightarrow f|F$ from A to $C(F)$ has an adjoint $T^*: M(F) \longrightarrow A^* = M(X)/A^\perp$ with $T^*\mu = \mu - A^\perp$, and T^* has closed range iff the induced 1-1 map $\overline{T^*}: M(F)/\ker T^* \longrightarrow M(X)/A^\perp$ has, and the last holds iff this 1-1 map has a bounded inverse by the open mapping theorem (along with the trivial fact that a bounded inverse guarantees a complete, and therefore closed, range). But $\ker T^* = (A|F)^\perp$ and thus $\overline{T^*}(\mu - (A|F)^\perp) = \mu - A^\perp$ for μ in $M(F)$, so $\overline{T^*}$ has a bounded inverse precisely when we have a c for which (2.1) holds.

COROLLARY 2.2. *$A|F = C(F)$ iff there is a $c \geqslant 1$ with*

(2.2) $$\|\mu_F\| \leqslant c\|\mu_{F'}\|, \quad \mu \in A^\perp.$$

PROOF. Suppose $A|F = C(F)$. Then $(A|F)^\perp = 0$ and (2.1) becomes $\|\nu\| \leqslant c\|\nu - A^\perp\|$, $\nu \in M(F)$; if we take $\nu = \mu_F$ for $\mu \in A^\perp$,

$$\|\mu_F\| \leqslant c\|\mu_F - A^\perp\| \leqslant c\|\mu_F - \mu\| = c\|\mu_{F'}\|.$$

On the other hand suppose (2.2) holds (with $c \geqslant 1$). Then $\mu \in (A|F)^\perp$ certainly implies $\mu = 0$, so $A|F$ is dense in $C(F)$, and we only have to see that is it closed, i.e. that (2.1) holds, or that

(2.3) $$\|\nu\| \leqslant c\|\nu - A^\perp\|, \quad \nu \in M(F).$$

But for $\mu \in A^\perp$, since $c \geqslant 1$,

$$\|\nu - \mu\| = \|(\nu - \mu)_F\| + \|(\nu - \mu)_{F'}\| \geqslant c^{-1}\|\nu - \mu_F\| + \|\mu_{F'}\|$$
$$\geqslant c^{-1}\|\nu\| - c^{-1}\|\mu_F\| + \|\mu_{F'}\| \geqslant c^{-1}\|\nu\|,$$

by hypothesis, which yields (2.3).

For any interpolation set F the isomorphism of $C(F) = A|F$ onto A/kF (where $kF = \{f \in A : f(F) = 0\}$) has precisely the least c satisfying (2.3) as its norm, as is easily verified. In any case we call F an *interpolation set of type* c if every $f \in C(F)$ has for $\epsilon > 0$ an extension in A of norm less than or equal to $(c + \epsilon)\|f\|$. As examples of those of type 1 we have those to which we can apply

COROLLARY 2.3. *$A|F = C(F)$ if $\mu_F = 0$, all $\mu \in A^\perp$.*

A set satisfying this last hypothesis is not only an interpolation set, as the corollary insures, but also a (generalized) peak set by 1.9, and is generally referred to as a peak interpolation set. (Conversely, a peak interpolation set satisfies the hypothesis of 2.3.) Since each of its points is a (generalized, which we will henceforth drop) peak set for $C(F) = A|F$, it is a peak set for A by 1.5, and $F \subset P$. The converse of this fact has only recently been obtained by Varopoulos, using a construction (employed earlier by Drury) from a related problem in harmonic analysis, which here boils down to a remarkably simple argument. This has (and no doubt will continue to have) further application to interpolation.

THEOREM 2.4. (VAROPOULOS). *If the compact interpolation set F lies in P, the set of peak points, then F is a peak interpolation set.*

As we know from 1.9, it suffices to show $\mu \in A^\perp$ implies $\mu_F \in A^\perp$. Fix such a μ and for $\epsilon > 0$ choose a compact $K \subset F'$ with $|\mu|(F' \backslash K) < \epsilon$. For each $x \in F$ we have an f in A peaking on a set containing x and lying in any preassigned neighborhood, e.g., K', of x. Replacing f by a power we thus have an f in A bounded by 1, with $f(x) = 1$ and $|f(K)| < \epsilon$, and so $|f(y) - 1| < \epsilon$ on a neighborhood V of x. By compactness then we have $f_1, \cdots, f_n$ in ball A and a measurable partition $F_1, \cdots, F_n$ of F for which $|f_i(x) - 1| < \epsilon$ on F_i, and $|f_i(K)| < \epsilon$. Within each F_i we choose a compact E_i so that $|\mu|(\bigcup(F_i \backslash E_i)) < \epsilon$.

Now if F is an interpolation set of type c then so is $\bigcup\{E_i : i = 1, \cdots, n\}$, and thus we can find $\beta_1, \cdots, \beta_n$ in A, $\|\beta_j\| \leqslant c + \epsilon$, with $\beta_j = \omega^{jk}$ on E_k, $k = 1, \cdots, n$, where $\omega = \exp(2\pi i/n)$. Then for the element $c_j = (n^{-1}\Sigma_{k=1}^n \omega^{-jk}\beta_k)^2$ of A one has, on E_r, $c_j = n^{-2}(\Sigma_{k=1}^n \omega^{-jk}\omega^{kr})^2 = \delta_{jr}$, so that $|\Sigma c_j f_j - 1| < \epsilon$ on each E_r, hence on the union, while

$$\sum_{j=1}^n |c_j| = n^{-2}\sum_j\left(\sum_k \omega^{-jk}\beta_k \sum_r \overline{\omega^{jr}\beta_r}\right)$$
$$= n^{-2}\sum_k\sum_r\left(\sum_j \omega^{(r-k)j}\,\beta_k\overline{\beta_r}\right)$$
$$= n^{-2}\sum_k n|\beta_k|^2 = n^{-1}\sum|\beta_k|^2 \leqslant (c + \epsilon)^2.$$

Thus $h = \Sigma_1^n c_j f_j \in A$ has norm less than or equal to $(c + \epsilon)^2$ and $|h - 1| < \epsilon$ on $\bigcup E_j$. Letting ϵ run over the sequence $\{1/n\}$ we get a sequence $\{h_n\}$ in A of norm less than or equal to $2c^2$ which approaches 1 a.e. $|\mu|$ on F and tends to 0 a.e. $|\mu|$ on F'. By dominated convergence then $h_n\mu \longrightarrow \mu_F$, and since $h_n\mu \perp A$, μ_F is also.

Evidently the above provides, in the presence of interpolation, a powerful means of constructing a bounded (approximate) partition of unity. Clearly it also proves

LEMMA 2.5. *Suppose $K_0, E_1, \cdots, E_n$ are disjoint compact subsets of an interpolation set E of type c, and $\eta > 0$. Then there are $c_1, \cdots, c_n$ in A with $\Sigma|c_j| \leq (c + \eta)^2$ and $c_j = \delta_{jk}$ on E_k, $c_j = 0$ on K_0. Moreover if there are $f_1, \cdots, f_n$ in A of norm less than M for which, for a compact $K \subset E'$ and $\epsilon > 0$, we have $|f_j(K)| < \epsilon$ and $|f_i - 1| < \epsilon$ on E_j, $j = 1, \cdots, n$, then $h = \Sigma c_j f_j \in A$ has $\|h\| < c^2 M$, $|h(K)| < c^2\epsilon$ and $|h - 1| < \epsilon$ on $\bigcup E_j$.*

Varopoulos has also applied this to answer the question of when the union of two sets of interpolation is one. In the case of the disc algebra $A = A(D)$, the sets of interpolation have been obtained by Carleson (see [H]): $F \subset D$ is an interpolation set if $F \cap T$ has linear measure 0 and $F \cap D^\circ$ is a sequence $\{z_n\}$ with

$$\prod_{j \neq k} \left| \frac{z_j - z_k}{1 - \bar{z}_j z_k} \right| \geq \delta > 0, \quad \text{for all } k. \tag{2.4}$$

Now it is easy to construct F_1 and F_2 for which (2.4) fails for $F_1 \cup F_2$, and precisely because pairs of elements (z_1, z_2), $z_i \in F_i$, can be arbitrarily close. Interpolation prevents this in general, in the guise of the open mapping theorem's guaranteeing that an interpolation set F is of type c for some c: for $x, y \in F$, $x \neq y$, we know there is an $f \in A$ of norm less than or equal to $(c + \epsilon)$ with $f(x) = 0, f(y) = 1$; thus the Gleason distance $\rho(x, y) = \frac{1}{2}\|x - y\|_{A^*}$ (half the distance between the evaluation functionals for x and y in A^*) is at least

$$\tfrac{1}{2} \sup \{|f(y)| : f \in \text{ball } A,\ f(x) = 0\} \geq 1/2c, \quad \text{for } x \neq y \text{ in } F.$$

THEOREM 2.6 (VAROPOULOS). *Suppose E_1, E_2 are sets of interpolation of type c_1, c_2 with $\inf \rho(e_1, e_2) = \delta > 0$ $(e_i \in E_i, e_1 \neq e_2)$. Then $E = E_1 \cup E_2$ is a set of interpolation (of type $\leq c_1^2 + c_2^2 + 2c_1^2 c_2^2 \delta^{-1}$, hence of type $\leq \delta^{-1}(c_1^2 + c_2^2 + 2c_1^2 c_2^2) = \delta^{-1}\psi(c_1, c_2)$ since $\delta \leq 1$).*

Note that the trivial inequality $|(z - w)/(1 - \bar{z}w)| \geq \frac{1}{2}|z - w|$ (or $2 \geq |1 - \bar{z}w|$) for z, w in D combines with the fact that for $f \in A$, $\|f\| < 1$, $g = (f - f(x))/(1 - \overline{f(x)}f)$ is in ball A to show that

$$\sup \{|g(y)| : g \in \text{ball } A,\ g(x) = 0\} \geq \tfrac{1}{2} \|x - y\|_{A^*} = \rho(x, y).$$

Thus $\rho(e_1, e_2) \geq \delta$ implies there is an f in ball A with $f(e_1) = 0$, $|f(e_2)| > \delta - \epsilon'$, and so an f in A with $f(e_1) = 0, f(e_2) = 1$, $\|f\| < \delta^{-1} + \epsilon$.

By 2.2, we want to show

(2.5) $$\|\mu_E\| \leq (c_1^2 + c_2^2 + 2c_1^2c_2^2\delta^{-1})\,\|\mu_{E'}\| \quad \text{for all } \mu \in A^\perp.$$

Fix μ again, and for $\epsilon > 0$ choose a compact $K \subset E_1 \backslash E_2$ with $|\mu|\,((E_1 \backslash E_2)\backslash K) < \epsilon$. For $x \in E_1$ and $y \in K$ there is an element f_{xy} of A of norm less than $\delta^{-1} + \epsilon$ with $f_{xy}(x) = 1, f_{xy}(y) = 0$, and thus we can obtain open coverings $\{U_1, \cdots, U_m\}$ of E_1 and $\{V_1, \cdots, V_n\}$ of K and elements $f_{ij} = f_{x_iy_j}$ of norm less than $\delta^{-1} + \epsilon$ in A with $|f_{ij}| < \epsilon$ on V_j and $|f_{ij} - 1| < \epsilon$ on $U_i, i = 1, \cdots, m, j = 1, \cdots, n$. Now within U_i and V_j we can find compact F_i and K_j for which $\bigcup F_i$ carries all but ϵ of the mass of $|\mu|$ on E_1 and $\bigcup K_j$ all but ϵ of the mass on K, with the $\{F_i\}$ pairwise disjoint and also the $\{K_j\}$ pairwise disjoint.

Applying 2.5 to $f_{1j}, f_{2j}, \cdots, f_{mj}$ and $F_1, \cdots, F_m$, we obtain a g_j in A of norm less than $c_1^2(\delta^{-1} + \epsilon)$ with $|g_j| < c_1^2\epsilon$ on K_j and $|1 - g_j| < \epsilon$ on $\bigcup F_i$. Applied again to $K_1, \cdots, K_n$ and $1 - g_1, \cdots, 1 - g_n$ the lemma now yields an f in A of norm less than $c_2^2(c_1^2(\delta^{-1} + \epsilon) + 1)$ with $|f| < c_2^2\epsilon$ on $\bigcup F_i$ and $|1 - f| < c_1^2$ on $\bigcup K_j$. Set $g = 1 - f$, and $\eta = (c_1^2 + c_2^2)\epsilon$ so $|g| < \eta$ on $\bigcup K_j$, $|1 - g| < \eta$ on $E_0 = \bigcup F_i$, and $\|g\| \leq c_2^2(c_1^2(\delta^{-1} + \epsilon) + 1) + 1 = M$

Noting that $g\mu \in A^\perp$, and that (2.2) applies to $F = E_1$ with $c = c_1$, we have

$$(1 - \eta)\,(\|\mu_{E_1}\| - \eta) \leq \|g\mu_{E_0}\| \leq \|g\mu_{E_1}\| \leq c_1\|g\mu_{E'_1}\|;$$

thus since $E'_1 = (E_1 \cup E_2)' \cup (E_2 \backslash E_1)$ and $\bigcup K_j$ carries all but 2ϵ of the mass of $\mu_{E_1 \backslash E_2}$, and $\eta > \epsilon$

$$(1 - \eta)\,(\|\mu_{E_1}\| - \eta) \leq c_1(\|g\mu_{(E_1 \cup E_2)'}\| + \|g\mu_{\cup K_j}\| + 2\epsilon M).$$

Now $|g| < \eta$ on $\bigcup K_j$, and $|g| \leq M$, so we have

$$(1 - \eta)\,(\|\mu_{E_1}\| - \eta) \leq c_1 M \|\mu_{(E_1 \cup E_2)'}\| + \eta\|\mu\| + 2\eta M,$$

and since $\eta = (c_1^2 + c_2^2)\epsilon > 0$ is arbitrary, and $M = c_2^2(c_1^2(\delta^{-1} + \epsilon) + 1)$.

(2.6) $$\|\mu_{E_1}\| \leq (c_2^2c_1^2\delta^{-1} + c_2^2)\,\|\mu_{(E_1 \cup E_2)'}\|.$$

By symmetry $\|\mu_{E_2}\| \leq (c_1^2c_2^2\delta^{-1} + c_1^2)\,\|\mu_{(E_1 \cup E_2)'}\|$, which with (2.6) yields (2.5).

Armed with Varopoulos' results, Allain Dufresnoy has quite recently succeeded in neatly characterizing $A|(E \cup F)$ when E, F are interpolation sets *and* $E \cap F$ is a peak set [Du].

THEOREM 2.7 (DUFRESNOY). *If E, F are two compact sets of interpolation for A and $E \cap F$ is a peak set then f is in $A|(E \cup F)$ iff $f \in C(E \cup F)$ and*

(2.7) $$|f(x) - f(y)| = o(\rho(x, y)), \quad x \in E, y \in F, x \neq y$$

(*i. e.*, $\rho(x, y)^{-1}\,|f(x) - f(y)| \longrightarrow 0$ *as* $\rho(x, y) \longrightarrow 0,\ x \in E, y \in F$).

COROLLARY 2.8 (DUFRESNOY). *With E, F as above, $A|(E \cup F)$ is self-adjoint and* Re *$A|(E \cup F)$ is a lattice.*

The last conclusion contrasts with the fact that $\operatorname{Re} A$ cannot be a lattice (under the ususl ordering of course) unless $A = C(X)$ (a result of Bernard; see §5); there are examples where $A|(E \cup F) \neq C(E \cup F)$ and 2.8 applies, for example two interpolation sets for $A(D)$; cf. (2.4).

Our proof of Dufresnoy's results is a reworking of his ideas due to Davie and Gamelin. We shall work with $\rho(x, F) = \inf \{\rho(x, y) : y \in F\}$ rather than $\rho(x, y)$, and to begin we should note that $x \longrightarrow \rho(x, F)$ is lower-semi-continuous (precisely because it is $\frac{1}{2}\|x - F\|_{A^*}$; if $x_\delta \longrightarrow x$ in X, and so as evaluation functionals w^* in A^*, and $\rho(x_\delta, f_\delta) = \rho(x_\delta, F), f_\delta \in F$, then passing to a subnet $f_\delta \longrightarrow f \in F$ w^* and so $\rho(x, f) = \frac{1}{2}\|x - f\|_{A^*} \leqslant \underline{\lim}\, \rho(x_\delta, f_\delta) = \underline{\lim}\, \rho(x_\delta, F))$. Thus $\{x : \rho(x, F) < c\}$ is an open set in X. And since $\rho(x, F) = 0$ implies $x \in F$ if A separates X, which we can assume without loss of generality, lower-semi-continuity implies $\inf \{\rho(x, F) : x \in K\} > 0$ for K compact, $K \cap F = \varnothing$.

LEMMA 2.9. *Suppose E and F are compact interpolation sets of type c_E, c_F. Then there is a constant k (depending only on c_E, c_F) with the property that for every f in $C(E)$ vanishing near $E \cap F$ and satisfying*

$$|f(x)| \leqslant M\,\rho(x, F), \quad x \in E, \tag{2.8}$$

there is an $h \in A$ with $h|E = f$, $h|F = 0$, $\|h\| \leqslant Mk$.

PROOF. Suppose $K \subset E$ is the compact support of f so $K \cap F = \varnothing$ and $\inf_K \rho(x, F) > 0$. For each x in K, $\{x\} \cup F$ is an interpolation set of type $\rho(x, F)^{-1}c$ where $c = \psi(c_E, c_F)$ as in Varopoulos' Theorem 2.6. Thus we have an $f_x \in A$ with $f_x = 0$ on $F, f_x(x) = f(x)$, and $\|f_x\| \leqslant c\,|f(x)|/\rho(x, F) \leqslant cM$ by (2.8). For $x \in E\backslash K$, $f_x \equiv 0$ has the same properties.

Now let K_0 be a compact neighborhood of $E \cap F$ in E near which $f = 0$. We want to show that f lies in the closure in $C(E)$ of those g in a fixed ball in A which vanish on $K_0 \cup F$, and of course it suffices to show that f lies in the weak closure. But given a measure μ on E, and $\epsilon > 0$, choose an open $E_0 \supset K_0$ in E on which $f = 0$ with $|\mu|(E_0 \backslash K_0) < \epsilon$, and then, refining a finite open covering of $E\backslash E_0$, we find closed disjoint $E_1, \cdots, E_n \subset E\backslash E_0$ and companion $f_i = f_{x_i}$ (as above) with $|\mu|\,(E\backslash \bigcup_{i=0}^n E_i) < \epsilon$ and $|f_i - f| < \epsilon$ on E_i, $i = 1, \cdots, n$. By Lemma 2.5 we have a Drury-Varopoulos approximate partition of unity $c_1, \cdots, c_n \in A$, $\Sigma_1^n |c_j| \leqslant c_E^2$, with $c_j = \delta_{jk}$ on E_k, $c_j = 0$ on K_0. So $g = \Sigma_1^n c_j f_j$ is of norm less than or equal to $c_E^2\, cM$, $|f - g| < \epsilon$ on $\bigcup_1^n E_k$, while $g = 0$ on $K_0 \cup F$, where $f = 0$ also. Hence, since $\|f\| \leqslant M$, $|\mu(f - g)| \leqslant \epsilon|\mu|(\bigcup_1^n E_k) + (c_E^2 cM + M)\,(|\mu|\,(E_0 \backslash K_0) + |\mu|\,(E\backslash \bigcup_0^n E_i)) < \epsilon\,\|\mu\| + (c_E^2 cM + M)2\epsilon$ so that f lies in the uniform closure of the $c_E^2\, cM$ ball of those $g \in A$ vanishing on $K_0 \cup F$. Since $\rho(x, F)$ is bounded away from zero on $E\backslash K_0$, for each $\epsilon > 0$ we can find such a g with $|f(x) - g(x)| < \epsilon\,\rho(x, F), x \in E$.

In particular, let $\epsilon = M/2$ and $k = 2c_E^2 c$: if $f \in C(E)$ vanishes near $E \cap F$ and $|f(x)| \leqslant M\,\rho(x, F), x \in E$, we have a g in the $k/2$-ball of A vanishing near $E \cap F$ (and on F) with $|f(x) - g(x)| \leqslant (M/2)\rho(x, f), x \in E$. As in the proof of the open mapping theorem, we now iterate: there is a g_1 in the $k/4$-ball of A vanishing near $E \cap F$ (and on F) with $|f(x) - g(x) - g_1(x)| < (M/4)\,\rho(x, F), x \in E$; continuing we get an element $g + g_1 + g_2 + \cdots = h$ in the k-ball of A which vanishes on F and coincides with f on E, as desired.

THEOREM 2.10. *Suppose* E *and* F *are closed interpolation sets and* $f \in C(E \cup F)$ *vanishes on* F *and satisfies*

$$|f(x)| \leqslant M\,\rho(x, F), \quad x \in E, \tag{2.9}$$

$$|f(x)| = o(\rho(x, F)), \quad x \in E. \tag{2.10}$$

Then for $\epsilon > 0$ *there is an* h *in* A *with* $h = f$ *on* $E \cup F$ *and* $\|h\| \leqslant k(M + \epsilon)$, *where* k *depends only on* c_E, c_F.

PROOF. By (2.10) for each $\epsilon > 0$ there is a neighborhood U of F with $|f(x)| < \epsilon\,\rho(x, F)$ for $x \in E \cap U$. Thus if we let $\epsilon_j = 2^{-j-1}\epsilon$ we can select open neighborhoods U_j of F in $E \cup F$ with $U_{j+1}^- \subset U_j$, and $|f(x)| < 2^{-j-1}\epsilon\,\rho(x, F), x \in E \cap U_j$. For $j \geqslant 2$ let $V_j = E \cap U_{j-1}$, an open set in E, and let $V_1 = E \setminus U_2^-, F_0 = \bigcap U_i \supset F$. Then $E \setminus F_0 \subset \bigcup_1^\infty V_j$, so since the σ-compact space $E \setminus F_0$ is paracompact we can find a partition of unity* $\{g_n\}$ subordinate to $\{V_n\}$: $0 \leqslant g_n \leqslant 1, g_n \in C(E), g_n \equiv 0$ off V_n, and $\Sigma_1^\infty g_n(x) = 1, x \in E \setminus F_0$.

By Lemma 2.9 there are h_j in A for which $h_j = fg_j$ on $E \cup F$, $\|h_j\| \leqslant 2^{-j}\epsilon k$ for $j \geqslant 2$, and $\|h_1\| \leqslant kM$. Thus $f = \Sigma h_j$ on $E \cup F$, and $\|\Sigma h_j\| \leqslant k(M + \epsilon)$.

Of course (2.9) is necessary (for some M) if $f \in A|(E \cup F)$ and $f(F) = 0$. Moreover (2.10) is also necessary if each point of $E \cap F$ is a peak point.

LEMMA 2.11. *Suppose* E *and* F *are closed and each point of* $E \cap F$ *is a peak set. Then* $f \in A$ *implies* $|f(x) - f(y)| = o(\rho(x, y)), x \in E, y \in F$; *moreover if* $f(F) = 0$, $|f(x)| = o(\rho(x, F)), x \in E$.

In particular this applies if $E \cap F$ is a peak interpolation set.

We consider only the second case, the proof being essentially the same in the first case. If the conclusion fails then there is a $c > 0$ and a net $\{x_\delta\}$ in $E \setminus F$ with $0 < \rho(x_\delta, F) \to 0$ and $|f(x_\delta)/\rho(x_\delta, F)| \geqslant c$. Choose $y_\delta \in F$ so that $\rho(x_\delta, y_\delta) < 2\rho(x_\delta, F)$. We can assume $x_\delta \to x$ in $E \cap F$, passing to a subnet, and since $\rho(x_\delta, y_\delta) \to 0, y_\delta \to x$ as well. Now the functional

$$L_\delta(g) = (g(x_\delta) - g(y_\delta))/\rho(x_\delta, y_\delta),$$

in A^* has norm less than or equal to 2 and if L is a w^* cluster point of $\{L_\delta\}$, L must be a point derivation at x since

$$L_\delta(gh) = \frac{gh(x_\delta) - gh(y_\delta)}{\rho(x_\delta, y_\delta)} = g(x_\delta)\frac{h(x_\delta) - h(y_\delta)}{\rho(x_\delta, y_\delta)} + h(y_\delta)\frac{g(x_\delta) - g(y_\delta)}{\rho(x_\delta, y_\delta)}$$

implies $L(gh) = g(x)L(h) + h(x)L(g)$. Since

$$|(f(x_\delta) - f(y_\delta))/\rho(x_\delta, F)| = |f(x_\delta)/\rho(x_\delta, F)| \geqslant c > 0$$

---

* We could argue more directly: $\{V_1\} \cup \{V_j \setminus U_{j+1}^- : j \geqslant 2\}$ is a locally finite—in fact locally 3-fold—covering of $E \setminus F_0$, and a directly constructible partition subordinate to it will do almost as well.

shows $L(f) \neq 0$, we have a non-zero point derivation at a peak point, which cannot happen [Br., p. 78].

Now (2.10) and (2.11) contain all the information necessary to obtain (2.7). Indeed if E and F are interpolation sets and $E \cap F$ a peak set, and $f \in A \mid (E \cup F)$ then (2.7) follows from (2.11); and if (2.7) holds for $f \in C(E \cup F)$ then since F is an interpolation set we have a g in A with $g = f$ on F, and, since (2.7) holds for g, it holds for $f_0 = f - g$, an element of $C(E \cup F)$ vanishing on F. Since (2.7) for f_0 asserts that for $\epsilon > 0$ we have a $\delta > 0$ for which $|f_0(x)| = |f_0(x) - f_0(y)| < \epsilon \rho(x, y)$, if $x \in E$, $y \in F$ and $\rho(x, y) < \delta$, $|f_0(x)| \leqslant \epsilon \rho(x, F)$ if $\rho(x, F) < \delta$, $x \in E$; so (2.10) holds for f_0, whence (2.9) also holds for some M, and (2.10) asserts $f_0 \in A \mid (E \cup F)$, and so $f \in A \mid (E \cup F)$.

3. THE ORTHOGONAL MEASURES

In order to use orthogonal measures to find, say, the peak interpolation sets of A, or to show a particular f in $C(X)$ lies in A, requires a fair amount of information on the rather large set $A^\perp$. At present our knowledge of $A^\perp$ is quite imperfect, but it does allow some applications.

Some elements of $A^\perp$ are easily produced from multiplicative measures: $M_\phi - M_\phi \subset A^\perp$, and $A_\phi M_\phi \subset A^\perp$, where $A_\phi = \{f \in A: \phi(f) = 0\}$. It should be noted immediately that these may all reduce to the zero measure: Cole, in his thesis, produced an algebra A for which the only representing measure on Δ for each $\phi \in \Delta$ is the point mass (i.e., each point of Δ is a peak point) while $A \neq C(\Delta)$; thus both the above subsets of $A^\perp$ are trivial. (Cole's construction is described in [Br]; for a simple easily visualized example with Δ a subset of the unit sphere in $\mathbf{C}^2$, see the recent preprint of Basener [Ba].)

Nevertheless in some cases, as for example, $A = R(K)\,|\,\partial K$, such measures yield a w^* dense subset of $A^\perp$, and thus suffice to determine which f in $C(X)$ lie in A. There are two central results which yield a decomposition of $A^\perp$ showing this, and after discussing these we shall turn to a description of a recent improved viewpoint due to Cole, and Seever and König.

The first result is an abstract version of the F. and M. Riesz theorem which evolved from the special case of dirichlet algebras (where it was handled by an argument of Helson and Lowdenslager, as observed by Bochner) through work of Hoffman, Lumer, Ahern and the author, to the general setting. To begin, we need a notion of absolute continuity of a measure relative to a set of measures.

DEFINITION. *For a set M of probability measures on X, a set $F \subset X$ is called M-null if $\lambda F = 0$ for all $\lambda \in M$. A measure μ is M-absolutely continuous ($\mu << M$) if $\mu_F = 0$ for all M-null F, and M-singular if $\mu = \mu_F$ for some M-null F.*

There is trivially a Lebesgue decomposition of any (complex finite) measure μ relative to $M: \mu = \mu_F + \mu_{F'}$ where F is M-null and $\mu_{F'} << M$. Indeed choose an M-null F for which $\|\mu_F\|$ is maximum; if E is M-null then so is $E \cup F$ and thus

$$\|\mu_{E \cup F}\| = \|\mu_F\| + \|(\mu_{F'})_E\| \leqslant \|\mu_F\|$$

implies $(\mu_{F'})_E = 0$, so $\mu_{F'} << M$.

Now fix a m.l.f. ϕ of A.

THEOREM 3.1. *Let* $\mu \in A^{\perp}$, *and let* $\mu = \mu_F + \mu_{F'}$ *be its Lebesgue decomposition relative to* $M = M_{\phi}$. *Then* $\mu_F, \mu_{F'} \in A^{\perp}$.

Our proof applies just as well to $M^{\perp}$ in place of $A^{\perp}$, where M is any A-module in $C(X)$. It depends on the following extension of a fundamental lemma of Forelli.

FORELLI'S LEMMA 3.2. *If* $F = \bigcup K_n$, K_n *compact* M_{ϕ}*-null, then there are* a_n *in* ball A *with* $a_n \longrightarrow 0$ *on* F *and* $a_n \longrightarrow 1$ *a. e.* λ, *all* λ *in* M_{ϕ}.

To see how this implies 3.1, note that by regularity, F there can be taken as a union, as in 3.2, and with the a_n there $\{x: \lim a_n \neq 1\}$ is M-null, so $|\mu_{F'}|$-null: thus $a_n \longrightarrow 1$ a.e. $|\mu_{F'}|$, $a_n \longrightarrow 0$ on F, so that $a_n \mu \longrightarrow \mu_{F'}$ in norm by dominated convergence, whence $\mu_{F'}$ (and so μ_F) is orthogonal to A since $a_n \mu$ is.

The proof of 3.2 depends on (1.3): for $u \in C_R(X)$,

$$\sup\{\operatorname{Re}\phi(a) : a \in A,\ \operatorname{Re} a \leqslant u\} = \inf\{\lambda(u) : \lambda \in M_{\phi}\}. \tag{3.1}$$

Indeed for n fixed the set $U = \{u \in C_R(X): u \leqslant -n\chi_{K_n}\}$ is (with the usual order) a directed set which, indexed by itself, forms a nondecreasing net. By regularity for any λ in M_{ϕ}, $\lim_U \lambda(u) = \lambda(-n\chi_{K_n}) = 0$. Thus the nondecreasing net of functions $f_u : \lambda \longrightarrow \lambda(u)$ on the w^* compact set M_{ϕ} converges uniformly to 0 by Dini's theorem. So we have a $u \in U$ for which $\inf\{\lambda(u) : \lambda \in M_{\phi}\} > -n^{-4}$, and by (3.1) there is an $f_n \in A$ with

$$\operatorname{Re} f_n \leqslant -n\chi_{K_n}, \quad \operatorname{Re}\phi(f_n) > -n^{-4}.$$

We set $a_n = \exp[f_n]\,\overline{\operatorname{sgn}}\,\exp[\phi(f_n)]$. Then $|a_n| = \exp[\operatorname{Re} f_n] \leqslant 1$, so $a_n \in$ ball A, and on K_n, $|a_n| \leqslant \exp[-n]$ so that $a_n \longrightarrow 0$ on F. And since $\phi(\exp[f_n]) = \exp[\phi(f_n)]$,

$$1 \geqslant \phi(a_n) = |\exp[\phi(f_n)]| = \exp[\operatorname{Re}\phi(f_n)] \geqslant e^{-n^{-4}} \geqslant 1 - n^{-4}$$

which implies, for any λ in M_{ϕ}, that

$$\begin{aligned}\int |a_n - 1|^2\, d\lambda &= \int |a_n|^2 + 1 - 2\operatorname{Re} a_n\, d\lambda = 1 + 1 - 2\operatorname{Re}\phi(a_n) \\ &\leqslant 1 + 1 - 2(1 - n^{-4}) = 2n^{-4},\end{aligned} \tag{3.2}$$

whence

$$\int \sum_1^{\infty} |a_n - a_{n+1}|\, d\lambda \leqslant \sum_1^{\infty} \Big(\int |a_n - a_{n+1}|^2\, d\lambda\Big)^{1/2} \leqslant \sum_1^{\infty} 4n^{-2}.$$

The last of course implies the sequence $\{a_n\} = \{a_1 + \Sigma_1^{n-1}(a_{j+1} - a_j)\}$ converges a.e. λ for each $\lambda \in M_{\phi}$, and necessarily to 1 by (3.2). Our proof of 3.2 is complete.

Now (3.1) allows us to write $\mu \in A^{\perp}$ as an infinite sum of mutually singular elements of $A^{\perp}$,

$$\mu = \sum_1^{\infty} \mu_n + \sigma \tag{3.3}$$

where $\mu_n << M_{\phi_n}$ and σ is M_{ϕ}-singular for all ϕ (so σ is called "completely singular"). One has only to choose ϕ_1, so that

$$\|\mu_{\phi_1}\| \geq \tfrac{1}{2} \sup \{\|\mu_\phi\| : \phi \in \Delta\}$$

where μ_ϕ denotes the M_ϕ-absolutely continuous component of μ, and setting $\mu_1 = \mu_{\phi_1}$, $\mu_1' = \mu - \mu_1$ (the M_{ϕ_1}-singular component), choose ϕ_2 so

$$\|(\mu_1')_{\phi_2}\| \geq \tfrac{1}{2} \sup \{\|(\mu_1')_\phi\| : \phi \in \Delta\}; \text{ then set } \mu_2 = (\mu_1')_{\phi_2}, \mu_2' = \mu - \mu_1 - \mu_2,$$

etc. Since the sequence of mutually singular restrictions μ_n of μ must have a norm convergent sum, we can set $\sigma = \mu - \Sigma_1^\infty \mu_n$, and of course $\|\mu_n\| \longrightarrow 0$ while σ is M_{ϕ_n}-singular for each n. But if σ were not completely singular, then, for some ϕ, $\|\sigma_\phi\| > 0$, whence $\|\sigma_\phi\| > 2\|\mu_n\|$ for some n, and since $\|(\mu_{n-1}')_\phi\| > \|\sigma_\phi\|$ we would have $\|(\mu_{n-1}')_\phi\| > 2\|\mu_n\| \geq \sup \{\|(\mu_{n-1}')_\psi\| : \psi \in \Delta\}$ a contradiction.

Actually, the decomposition in (3.3) is unique and canonical, with μ_n the M_{ϕ_n}-a.c. component of μ, but (3.3) will suffice for the moment. An obvious question at this point is just that of the composition of $(A^\perp)_\phi = \{\mu \in A^\perp : \mu << M_\phi\}$. For a fixed λ in M_ϕ the elements of $L^2(\lambda)$ orthogonal to A are easily described. If $H^2(\lambda)$ is the closure of A in $L^2(\lambda)$, and $H_0^2(\lambda)$ the subset of elements of mean 0, then

$$L^2(\lambda) = H^2(\lambda) \oplus \overline{H_0^2(\lambda)} \oplus (\mathrm{Re}\, A)_{\lambda,2}^\perp \tag{3.4}$$

where $(\mathrm{Re}\, A)_{\lambda,2}^\perp$, the subspace of elements orthogonal to Re A, also clearly coincides with the (complex) span of the real valued elements of $L^2(\lambda)$ orthogonal to A.

Indeed the fact that $H^2(\lambda)$ and $\overline{H_0^2(\lambda)}$ are orthogonal in $L^2(\lambda)$ is just the fact that the multiplicativity of λ on A extends to $H^2(\lambda)$: if $a_n \longrightarrow f$ in $L^2(\lambda)$, $a_n \in A$, then $\int fa\,d\lambda = \lim \int a_n a\,d\lambda = \lim \int a_n d\lambda \int a\,d\lambda = \int f d\lambda \int a\,d\lambda$, so if $b_n \longrightarrow g$, $\int fg\,d\lambda =$ $\lim \int f b_n d\lambda = \lim \int f d\lambda \int b_n d\lambda = \int f d\lambda \int g\,d\lambda$. And, if $f \in L^2(\lambda) \ominus (H^2(\lambda) \oplus \overline{H_0^2(\lambda)})$ evidently f is orthogonal to $r + a + \bar{a}$ for each $a \in A$ with $\phi(a) = 0 = \lambda(a)$, and thus to Re A.

Now for $f \in C(X)$ to be orthogonal to the orthogonal measures (3.4) provides amounts to f lying in $H^2(\lambda)$: thus f is orthogonal to the w^* closed span of the measures $(H_0^2(\lambda) \oplus (\mathrm{Re}\, A)_{\lambda,2}^\perp)\lambda$, all λ in M_ϕ, precisely when $f \in \bigcap\{H^2(\lambda) : \lambda \in M_\phi\}$. As we shall see in a moment the last is equivalent to f being orthogonal to $(A^\perp)_\phi$.

THEOREM 3.3. *If* $f \in C(X)$ lies in $\bigcap\{H^2(\lambda) : \lambda \in M_\phi\}$ *then there is a sequence* $\{a_n\}$ *in* A *for which*

$$\sup_{M_\phi} \int |f - a_n|^2 d\lambda \longrightarrow 0. \tag{3.5}$$

I. e., if f *is approximable in each individual* $L^2(\lambda)$, *it is simultaneously approximable in all.*

THEOREM 3.4 (HOFFMAN-WERMER). *If* (3.5) *holds for any bounded Borel* f *then there is a sequence* $\{b_n\}$ *in the* $\|f\|_\infty$*-ball of* A *for which* $b_n \longrightarrow f$ *a. e.* λ, *all* λ *in* M_ϕ.

The Hoffman-Wermer theorem follows by an argument quite similar to that in the Forelli lemma. Indeed take $\|f\|_\infty = 1$, and suppose the left side of (3.5) is $< n^{-4}$ for each n. Then $a_n \to f$ a.e. λ, all λ in M_ϕ as before. For $\epsilon > 0$, set $G_n = \{x: |a_n(x)| > 1 + \epsilon\}$. Then $\sup_{M_\phi} \lambda G_n \to 0$ clearly and

$$\int_{G_n} \log |a_n| d\lambda \leqslant \int_{G_n} |a_n| d\lambda \leqslant \int_{G_n} |a_n - f| d\lambda + \int_{G_n} |f| d\lambda$$
$$\leqslant \sup \left(\int |a_n - f|^2 d\lambda \right)^{1/2} + \sup \lambda G_n$$

so $\eta_n = \sup_\lambda \int_{G_n} \log |a_n| d\lambda \to 0$. Since

$$\int \log^+ |a_n| d\lambda \leqslant \int_{G_n} \log |a_n| d\lambda + \log(1 + \epsilon) \leqslant \eta_n + 2\epsilon \text{ for } n > n_\epsilon$$

for each λ and $\epsilon > 0$, $\eta'_n = \sup_\lambda \int \log^+ |a_n| d\lambda \to 0$.

Now set

$$u_n = -\log^+ |a_n| \in C_R(X). \tag{3.6}$$

We have $-\eta'_n = \inf \{\lambda(u_n) : \lambda \in M_\phi\} \to 0$ so that by (3.1) again we have f_n in A with $\operatorname{Re} f_n \leqslant u_n \leqslant 0$, $\operatorname{Re} \phi(f_n) > -\eta'_n - (1/n) \to 0$. Choose n_j, so that $\operatorname{Re}(f_{n_j}) > -j^{-4}$, and set $g_j = \exp [f_{n_j}] \overline{\operatorname{sgn}} \exp [\phi(f_{n_j})]$. Then $|a_{n_j} g_j| = |a_{n_j}| \exp [\operatorname{Re} f_{n_j}] \leqslant |a_{n_j}| \exp [u_{n_j}] \leqslant 1$ by (3.6) and $|g_j| = \exp [\operatorname{Re} f_{n_j}] \leqslant 1$ since $f_{n_j} \leqslant u_{n_j} \leqslant 0$. Since

$$1 \geqslant \phi(g_j) = |\exp [\phi(f_{n_j})]| = \exp [\operatorname{Re} \phi(f_{n_j})] > 1 - j^{-4}$$

again leads to the fact that $g_j \to 1$ a.e. λ, all λ in M_ϕ, we conclude that $b = a_{n_j} g_j \to f$ a.e. λ for each λ in M_ϕ as desired.

Turning to 3.3, note that its hypothesis can be stated as

$$\sup_{\lambda \in M_\phi} \inf_{a \in A} \int |f - a|^2 d\lambda = 0$$

while its conclusion asserts that

$$\inf_{a \in A} \sup_{\lambda \in M_\phi} \int |f - a|^2 d\lambda = 0;$$

thus the result follows from the following version of von Neumann's minimax theorem of game theory.

THEOREM 3.5. *Suppose A, M are convex subsets of topological vector spaces, M is compact, and F is a non-negative function on $A \times M$ such that $a \to F(a, m)$ is convex while $m \to F(a, m)$ is concave and continuous. Then*

$$\inf_a \sup_m F(a, m) = \sup_m \inf_a F(a, m). \tag{3.7}$$

Trivially one has

$$0 \leqslant \sup_m \inf_a F(a, m) \leqslant \inf_a \sup_m F(a, m) < \infty,$$

the last since M is compact and $m \to F(a, m)$ continuous. Let us replace F by $F - \sup_m \inf_a F$, so F is now only real valued but $\sup \inf F(a, m) = 0$. If (3.7) fails then

$$0 = \sup_m \inf_a F(a, m) < c < \inf_a \sup_m F(a, m).$$

For each a there is an m with $F(a, m) > c$ and thus if $c_i \geqslant 0$, $\Sigma c_i = 1$, $a_i \in A$, there is an m with $c < F(\Sigma c_i a_i, m)$, so $c < \Sigma c_i F(a_i, m)$ since $a \to F(a, m)$ is convex. Hence the closed convex hull K of $\{F(a, \cdot): a \in A\}$ in $C_R(M)$ is at a distance at least c from the non-positive cone, and we have a measure μ of norm 1, $\leqslant 0$ on the non-positive cone and $\geqslant \epsilon > 0$ on K; in other words we have a probability measure μ for which

$$0 < \epsilon \leqslant \int F(a, m)\mu(dm) \quad \text{for all } a \in A.$$

But $\int F(a, m)\mu(dm) \leqslant F(a, \int m\mu(dm))$, so

$$0 < \epsilon \leqslant \inf_a F\left(a, \int m\mu(dm)\right) \leqslant \sup_m \inf_n F(a, m) = 0,$$

our contradiction.

Theorems 3.3 and 3.4 combine to show any $f \in \bigcap_{\lambda \in M} H^2(\lambda) \cap C(X)$ is necessarily orthogonal to $(A^\perp)_\phi = \{\mu \in A^\perp: \mu << M_\phi\}$. For they yield a sequence b_n in the $\|f\|_\infty$ ball of A for which $\{x: \lim b_n(x) \neq f(x)\}$ is M_ϕ-null, hence $|\mu|$-null for μ in $(A^\perp)_\phi$. Thus $\mu(f) = \lim \mu(b_n) = 0$ by dominated convergence. Indeed

THEOREM 3.6. *Suppose* (ball $A^\perp)^e$ *has no completely singular elements. Then*

$$f \in \bigcap_{\lambda \in M_\phi} H^2(\lambda) \cap C(X)$$

for all ϕ *in* Δ *implies* $f \in A$.

For any element μ of the set (ball $A^\perp)^e$ of extreme points must be $<< M_\phi$ for some ϕ by (3.3) (which must contain exactly one term). Since $\mu(f) = 0$ then, we know $f \in A$ by Kreĭn-Milman.

THEOREM 3.7. *Suppose we have closed subalgebras* $A \subset B \subset C(X)$ *and*

(i) (ball $A^\perp)^e$ *has no completely singular elements,*

(ii) $\Delta_A = \Delta_B$,

(iii) $(\text{Re } A)^\perp = (\text{Re } B)^\perp$ (i.e. Re A *and* Re B *have the same uniform closures*).

Then $A = B$.

Of course all but (i) are necessary as well. The meaning of (ii) is just that a m.l.f. ϕ on A is the restriction to A of a unique element of Δ_B. Actually along with (iii) this implies A and B share the same multiplicative measures: for clearly $M_\phi(B) \subset M_\phi(A)$ (i.e., any λ representing ϕ on B does so on A); on the other hand if $\lambda' \in M_\phi(A)$ then $\lambda' - \lambda \perp \text{Re } A$, hence $\perp \text{Re } B$, and therefore $\perp B$, so both represent the same functional on B and $\lambda' \in M_\phi(B)$.

Now for $\lambda \in M_\phi(A) = M_\phi(B)$ we have, with obvious notation,

$$\begin{array}{ccccccc} L^2(\lambda) = & H^2(A, \lambda) & \oplus & \overline{H_0^2(A, \lambda)} & \oplus & (\text{Re } A)_{2,\lambda}^\perp, \\ & \cap & & \cap & & \| \\ L^2(\lambda) = & H^2(B, \lambda) & \oplus & \overline{H_0^2(B, \lambda)} & \oplus & (\text{Re } B)_{2,\lambda}^\perp; \end{array}$$

reading between the lines we conclude from orthogonality that $H^2(A, \lambda) = H^2(B, \lambda)$, and evidently our result now follows from 3.6.

In particular, 3.7 applies to rational approximation in **C**; indeed for $A = R(K)$ we can easily identify Δ_A with K, and for $B = A(K)$ the same is true but not so simple (see Browder's book) while there are no completely singular measures at all, by a result of Wilken (loc. cit.). So 3.7 (with $X = \partial K$ for economy, although it amounts to the same thing with $X = K$ if we avoid measures) simply asserts $R(K) = A(K)$ if $\operatorname{Re} R(K)$ and $\operatorname{Re} A(K)$ have the same uniform closure. Rather remarkably, a strictly non-self-adjoint problem reduces to a self-adjoint one. (In fact the problem is just whether $A(K)$ and $R(K)$ have the same multiplicative measures [Gar-Gl].)

An alternative approach to the abstract F. and M. Riesz theorem was given by König and Seever [K–S]; it produced an a priori distinct decomposition of $\mu \in A^\perp$ as follows: choose $\lambda \in M_\phi$ for which $\|\mu_\lambda\|$ is a maximum, where μ_λ is the component of μ absolutely continuous with respect to λ in the usual Lebesgue decomposition. Their result then asserts $\mu_\lambda \perp A$. In fact μ_λ coincides with our M_ϕ-a.c. component of μ, which, as was pointed out by Rainwater [R], is a consequence of minimax. Indeed $\nu = \mu - \mu_\lambda$ is singular with respect to all $\lambda' \in M_\phi$, and if one shows it is (therefore) carried by an M_ϕ-null set F, then since μ_λ is clearly M_ϕ-a.c., we evidently will have the same decomposition as earlier. But singularity of $|\nu|$ and λ' imply

$$\inf_F (|\nu|(f) + \lambda'(1-f)) = 0$$

where $F = \{f \in C_R(X) : 0 \leq f \leq 1\}$; thus

$$0 = \sup_{M_\phi} \inf_F (|\nu|(f) + \lambda'(1-f)) = \inf_F \sup_{M_\phi} (|\nu|(f) + \lambda'(1-f)),$$

by 3.5, so there are f_n in F with

$$|\nu|(f_n) + \lambda'(1-f_n) < n^{-2} \quad \text{for all } \lambda',$$

whence clearly $f_n \longrightarrow 0$ a.e. $|\nu|$, $f_n \longrightarrow 1$ a.e. λ', for each λ', and $\{x : \lim f_n(x) = 0\}$ is an M_ϕ-null set carrying ν.

The main device used by König and Seever was the notion of a reducing band of measures. A *band of measures* S (on any measurable space) is simply a (norm closed) linear space of measures with the property that $\nu << |\mu|$, $\mu \in S$, imply $\nu \in S$. (So it contains $L_1(\mu)$, or more precisely we should write $L_1(|\mu|)|\mu|$, for each of its elements μ; we shall want to identify $L_1(\mu)$ and $L_1(\mu)\mu$, the corresponding measures, occasionally.) Given a sub-band S of a band B one has a complementary band S' relative to B consisting of all μ in B singular to all elements of S, and $B = S \oplus S'$ (via the ordinary Lebesgue decomposition; for $\mu \in B$ choose λ in S with $\|\mu_\lambda\|$ a maximum–using an infinite convex combination of approximately maximizing λ makes this possible–and then $\mu - \mu_\lambda \in S'$). Now a *reducing* band in B is a sub-band S for which $\mu \in A^\perp \cap B$, $\mu = \mu_1 + \mu_2$, $\mu_1 \in S$, $\mu_2 \in S'$, imply $\mu_i \in A^\perp$. (S can be a trivial reducing band in the sense that all such decompositions are trivial, i.e. $A^\perp \cap B \subset S$, or $A^\perp \cap B \subset S'$, of course.)

This notion has quite recently been incorporated into the most general approach to decomposing orthogonal measures by Brian Cole–an approach in which (for $B = M(X)$) the decomposition is accomplished via idempotents in A^{**}. Cole has yet to write up his results, but here is a brief summary. [1)]

For a band B on a measurable space X and an algebra A of bounded measurable functions containing the constants, set $L^\infty(B) = B^*$ (just the dual of the Banach space B) and $H^\infty(B) = (B/(A^\perp \cap B))^*$, just the w^* closed span of A in B^*. (When $B = M(X)$ since $M(X)/A^\perp = A^*$, $H^\infty(B) = A^{**}$, a subspace of $L^\infty(M(X)) = M(X)^*$.) Now $L^\infty(B)$ and its subspace $H^\infty(B)$ are each function algebras, and indeed are inverse limits: $L^\infty(B) = \varprojlim L^\infty(\mu)$, $H^\infty(B) = \varprojlim H^\infty(\mu)$, so the multiplication is apparent. (Here we write $L^\infty(\mu)$ for $L^\infty(|\mu|)$ and $H^\infty(\mu)$ is the w^* closure of A in $L^\infty(\mu)$.) For an element $\rho = \{\rho_\mu\}$ of $L^\infty(B) = \varprojlim L^\infty(\mu)$ we can define multiplication of elements of B by ρ, $\rho \cdot \mu = \rho_\mu \mu$, so B is an $L^\infty(B)$ module.

Now a sub-band S of B is simply ρB for some idempotent in $L^\infty(B)$, while S is a reducing band if $\rho \in H^\infty(B)$. Moreover S is a minimal reducing band (i.e., not non-trivially expressible as $S_1 \oplus S_2$ with S_i reducing bands) if ρ is a minimal idempotent in $H^\infty(B)$ (there is an obvious partial order among the real elements of $L^\infty(B)$—those with each ρ_μ real valued, and minimality for our idempotent is relative to this order). The fundamental F. and M. Riesz theorem is now that if ϕ is a w^* continuous homomorphism on $H^\infty(B)$ and $M_\phi(B)$ is the set of probability measures in B representing ϕ, then $S_\phi = M_\phi(B)''$, the sub-band generated by $M_\phi(B)$, is a minimal reducing band. (For $B = M(X)$ in our old context, ϕ a m.l.f. on A has a w^* continuous extension to $H^\infty(M(X)) = A^{**}$, and $M_\phi(M(X)) = M_\phi$. Thus $S_\phi \cap A^\perp$ is precisely $(A^\perp)_\phi$.) [2)]

In fact Cole obtains a grand reducing decomposition

$$B = \sum B_\alpha \oplus C \oplus S$$

where the B_α are minimal (non-trivial) reducing bands, S is singular with respect to all elements of $A^\perp \cap B$, and C is the sub-band of B generated by all μ in $A^\perp \cap B$ which are totally decomposable in the sense that for $\epsilon > 0$ there are orthogonal idempotents $\rho_1, \rho_2, \cdots, \rho_n \in H^\infty(B)$ with $\mu = \Sigma_1^n \rho_i\mu$ and $\|\rho_i\mu\| < \epsilon$. In particular, for the bidisc algebra $A(D^2)$, Cole shows this decomposition can be written (for $X = T^2$, the Silov boundary)

$$M(T^2) = \sum B_\alpha \oplus C_1 \oplus C_2 \oplus S$$

where each $B_\alpha = M''_{\phi_\alpha}$, with one for one (arbitrary) point ϕ_α in each non-trivial Gleason part, i.e., one term for $(D^2)^\circ$, and one for each open disc $\{e^{i\theta}\} \times D^\circ$ or $D^\circ \times \{e^{i\theta}\}$ in the topological boundary, S is as before, and C_1 consists of all measures carried by sets $E \times T$, with $|E| = 0$, which vanish on sets $\{e^{i\theta}\} \times T$; C_2 is defined similarly. As a consequence, (ball $A(D^2)^\perp)^e$ has no completely singular elements.

---

1) After the conference I learned that Seever had noted a somewhat similar approach, obtaining (up to a point I shall identify below) rather parallel results. These appear in *Algebras of continuous functions on hyperstonian spaces* (preprint).

2) Beyond this point, the results of Seever diverge from Cole's.

Aside from the basic idea of using A^{**} to provide the decomposition, the main tool in seeing that $M_\phi(B)''$ is a reducing band is a theorem of Hoffman and Rossi which depends on the Kreĭn-Smulian theorem (which thus assumes the functional analytic role of Hahn-Banach (and more) in what we have done here). Finally it should be mentioned that Cole's argument yields an F. and M. Riesz theorem quite independent of the notion of a band.

THEOREM. *Let* m *be a probability measure for which* $M_\phi(m) = M_\phi \cap L^1(m) \neq \emptyset$. *If* $\mu \in A^\perp \cap L^1(m)$ *and we choose* λ *in* $M_\phi(m)$ *for which* $\|\mu_\lambda\|$ *is a maximum, then* $\mu_\lambda \perp A$.

This of course includes 3.1 (modulo Rainwater's result) since for any $\mu \in A^\perp$ we can choose λ_0 in M_ϕ so that $\|\mu_{\lambda_0}\|$ is a maximum, and then let $m = |\mu| + \lambda_0$; trivially the λ in $M_\phi(m)$ yielding the maximum in the theorem is λ_0.

(We have barely touched upon the Hardy spaces, in our introduction of $H^2(\lambda)$; there is a well developed theory to be found in [Gam] and in the broadest context, in Lumer's notes [L 1]. For further applications of F. and M. Riesz, see [Gl 1], and for some applications of minimax, see [K], [L 2].)

4. RATIONAL APPROXIMATION

The application to rational approximation of orthogonal measures along the lines of the preceding section is hampered by the general abundance of multiplicative measures on $R(K)$ (even, as usual, restricted to ∂K for economy). For some time it has been known that there is a single measure on ∂K, an infinite convex combination of multiplicative measures, which determines when an f in $A(K)$ lies in $R(K)$ [Gar–Gl], but one usually has no hope of finding it. Quite recently, however, Davie has obtained a result [D] which has allowed Gamelin and Garnett [Ga–Gar] to prove that, for any compact $K \subset \mathbf{C}$,

$$R(K) = C(K) \cap H^\infty(\lambda_Q)$$

where λ_Q is the plane area measure restricted to Q, the non-peak points in K of $R(K)$, and $H^\infty(\lambda_Q)$ is the w^* closure of $R(K)$ in $L^\infty(\lambda_Q)$: thus $f \in C(K)$ lies in $R(K)$ if there is a bounded sequence in $R(K)$ converging a. e. λ_Q to f.

While the question of when $R(K) = A(K)$ seems at present to be best handled by the classical techniques of Vituskin, there are a variety of other questions to which a combination of both classical and functional analytic tools have been remarkably well applied, principally, by Davie, and Gamelin and Garnett (see bibliography) e. g., as in determining when $R(K)$ (or $A(K)$) is dirichlet on ∂K (i. e., has no real orthogonal measures on ∂K). This turns out to be naturally related to questions of "pointwise bounded approximation," for example, whether each element of $H^\infty(K^\circ)$, the bounded analytic functions on K°, is a pointwise limit of a bounded sequence in $R(K)$ or $A(K)$. (In fact, if $R(K)$ or $A(K)$ is dirichlet on ∂K each component of K° must be simply connected [Gam] and when each component is, Gamelin and Garnett have shown $A(K) | \partial K$ is dirichlet iff $A(K)$ is pointwise boundedly dense in $H^\infty(K^\circ)$, while $R(K) | \partial K$ is dirichlet iff $R(\partial K) = C(\partial K)$ and $R(K)$ pointwise boundedly dense in $H^\infty(K^\circ)$.) So far all complete answers in this area seem to be connected with the notions of analytic and continuous analytic capacity, but for our present purposes we can ignore these notions entirely.

Briefly—quite briefly—the Vituskin approach splits an element of $A(K)$ into a sum of functions (extendable to be) analytic on all but a small subset of the Riemann sphere S via a partition of unity and an integral transform, and then utilizes analytic capacity conditions to approximate each piece by a function analytic near K in such a way that interference among the various approximations is controlled. We shall need only a bit of the first part of this approach, which we shall develop later, to discuss Davie's result.

For an *open set* U in $\mathbf{C}$, $A(U)$ denotes the continuous functions on U^- analytic on U, and $H^\infty(U)$ all bounded analytic functions on U.

THEOREM 4.1 (DAVIE). *If $f \in H^\infty(U)$ is the pointwise limit (on U) of a bounded sequence in $A(U)$ there is a sequence $\{f_n\}$ in $A(U)$ with $\|f_n\|_\infty \leqslant \|f\|_\infty$ converging to f pointwise on U.*

As an immediate consequence the set B of all such pointwise limits of bounded sequences is uniformly closed. (Pointwise convergence of a bounded sequence is uniform convergence on compacta, so a uniform bound, and the diagonal process, insure this uniform closure.) For any subset A of $H^\infty(U)$ one could consider the set B_A of pointwise limits of bounded sequences in A, of course; for A the polynomials, B_A is *not* uniformly closed, as Rubel and Shields [R–S] showed by an example.

Davie's proof divides into two parts, the first of which involves the functional analysis.

STEP I. *We shall assume the theorem false and, for each $\epsilon > 0$, show there is a sequence $\{g_n\}$ in* ball $A(U)$, $F \subset U^-$ *and a measure σ in $A(U)^\perp$ with $g_n \longrightarrow g$ pointwise on U, where $\|g\| < \epsilon$, $|1 - g_n| < \epsilon$ on F, and $\sigma(F) \neq 0$.*

If the theorem is false, we can find an f in B of norm 1 for which $1 < c = \inf\{\sup \|f_n\| : \{f_n\} \subset A(U), f_n \longrightarrow f$ pointwise on $U\}$. We can, of course, find one such sequence $\{f_n\}$ with $\|f_n\| \leqslant c$, which we fix. Choose m so large that $c^{-m} < \epsilon$, and let $\eta = \frac{1}{2} c\delta$, where $|1 - z| < \delta$ implies $|1 - z^m| < \epsilon$, $\delta > 0$. Now from the definition of c, we can find a compact $K \subset U$ for which f is not the uniform limit on K of elements of the $(c - \eta)$-ball of $A(U)$ (otherwise the diagonal process would yield f a pointwise limit of a sequence from the $(c - \eta)$-ball). Thus by the separation theorem for convex sets we have a measure μ on K with $\mu(f) > 1 \geqslant |\mu(h)|$, for all h in the $(c - \eta)$-ball of $A(U)$. As a functional on $A(U)$, $h \longrightarrow \mu(h)$ is then of norm $\leqslant (c - \eta)^{-1}$, so there is a measure ν on U^- of norm $\leqslant (c - \eta)^{-1}$ for which $\mu(h) = \nu(h)$, $h \in A(U)$, i.e., with $\sigma = \mu - \nu \in A(U)^\perp$. We can of course take $c - \eta > 1$.

Recall our fixed sequence $\{f_n\}$ in the c-ball of $A(U)$ converging pointwise to f on U. If G is any w^* cluster point of $\{f_n\}$ in $L^\infty(|\mu| + |\sigma|)$, so that $\sigma(f_n) = 0$ implies $\sigma(G) = 0$, we have

$$1 < \mu(f) = \lim \mu(f_n) = \mu(G) = \nu(G) \leqslant \|G\| \, \|\nu\| \leqslant \|G\| \, (c - \eta)^{-1}$$

and $\|G\| > c - \eta$. Evidently $\|G\| \leqslant c$ and $|G| \leqslant 1$ a.e. $|\mu|$ since $f_n \longrightarrow f$ on K and $\|f\|_\infty = 1$. So since $|G| > c - \eta$ on a set of positive $|\sigma| + |\mu|$-measure, $c \geqslant |G| > c - \eta$ on a set F_1 with $|\sigma| \, (F_1) > 0$. Finally, since the annulus $\{z : c - \eta < |z| \leqslant c\}$ can be covered by finitely many discs of radius 2η and centers on $|z| = c$, the inverse under G of one such disc has positive $|\sigma|$-measure, and multiplying all our functions by $e^{i\theta}$ for some θ we can assume that disc is centered at c. So we can find a compact $F_2 \subset F_1$ with $|\sigma| \, (F_2) > 0$ and

$$|c - G| < 2\eta \text{ on } F_2. \tag{4.1}$$

Now passing to a subsequence we can assume $f_n \longrightarrow G$ w^* in $L^\infty(|\sigma|)$, hence weakly in $L^2(|\sigma|)$. Thus we get a sequence of convex combinations $\{f_n'\}$ of the f_n

which converge to G in the norm of $L^2(|\sigma|)$, which still converge to f pointwise in U, and, of course, are still bounded by c. Again passing to a subsequence we can assume $f_n' \longrightarrow G$ a.e. $|\sigma|$, and by Egoroff we can find a compact $F \subset F_2$ with $\sigma F \neq 0$ on which $f_n' \longrightarrow G$ uniformly, so $|f_n' - c| < 2\eta$ for $n \geqslant n_0$, by (4.1).

We set $g_n = (c^{-1}f_n')^m$ and $g = (c^{-1}f)^m$. Then $g_n \in A(U), g_n \longrightarrow g$ pointwise on U, while $\|g\|_U < \epsilon$ since $c^{-m} < \epsilon$ and $|f| \leqslant 1$. Moreover on F, $|1 - g_n| = |1 - (c^{-1}f_n')^m| < \epsilon$ by our choice of η since $|z - c| < 2\eta = c\delta$ implies $|1 - c^{-1}z| < \delta$. Thus Step I is complete.

In Step II we will show we cannot have the conclusion in Step I if $\epsilon > 0$ is sufficiently small; our tools come from the initial steps of the Vituskin approach along with a subsidiary construction due to Øksendal. At various points we will use $A_1, A_2, \cdots$ to denote absolute constants. Our first task is the construction of certain partitions of unity on $\mathbf{C}$.

We begin with a single non-negative $C^{(1)}$ function ψ, supported on the square S_0 defined by $|\mathrm{Re}\, z| \leqslant \frac{1}{2}, |\mathrm{Im}\, z| \leqslant \frac{1}{2}$, with

$$\iint \psi\, dxdy = 1 \quad \text{and} \quad |\mathrm{grad}\, \psi| = \left(\left(\frac{\partial\psi}{\partial x}\right)^2 + \left(\frac{\partial\psi}{\partial y}\right)^2\right)^{1/2} \leqslant 16.$$

(For example, we can take $\psi(x, y) = 4t(x)t(y)$ where t is a slightly smoothed version of the usual triangular function based on $[-\frac{1}{2}, \frac{1}{2}]$ with $t(0) = 1$, so $|t'| \leqslant 2 + \epsilon$ and $|\partial\psi/\partial x| \leqslant 4(2 + \epsilon)$.) If we let $\{S_k\}$ be a partition of $\mathbf{C}$ into squares of side one obtained by translating S_0 about appropriately, then

$$\psi_k(\zeta) = \chi_{S_k} * \psi(\zeta) = \iint_{S_k} \psi(\zeta - z)\, dx\, dy \geqslant 0 \quad (z = x + iy)$$

has support in $S_0 + S_k$, a square of side 2, is of course $C^{(1)}$, with $|\mathrm{grad}\, \psi_k| \leqslant 16$, and

$$\sum \psi_k(\zeta) = \iint_{\mathbf{C}} \psi(\zeta - z)\, dx\, dy = 1, \quad \text{all } \zeta,$$

so that $\{\psi_k\}$ provides a partition of unity. Evidently each ψ_k has compact support in a disc of radius 2, and no more than (say) 25 such discs meet any given point. Now for $\delta > 0$ we set

$$\phi_k(z) = \phi_{k\delta}(z) = \psi_k(2z/\delta)$$

so that we obtain a partition of unity with each ϕ_k supported by a disc D_k of radius δ, and

$$|\mathrm{grad}\, \phi_k| = \left(\left(\frac{\partial\phi_k}{\partial x}\right)^2 + \left(\frac{\partial\phi_k}{\partial y}\right)^2\right)^{1/2} \leqslant 2 \cdot 16/\delta = 32\delta^{-1}.$$

Let $\{\phi_1, \cdots, \phi_r\}$ be the subset of $\{\phi_k\}$ supported by discs meeting our set F obtained in Step I, so that $\Sigma\phi_i = 1$ near F. We define

$$\text{(4.2)} \qquad h_i^{(n)}(\zeta) = \frac{1}{\pi} \iint_U \frac{\partial\phi_i}{\partial\bar{z}} \frac{g_n(z)}{z - \zeta}\, dx\, dy, \quad \zeta \in \mathbf{C} \ (z = x + iy)$$

(which exists everywhere and is continuous as the convolution of the locally integrable function $1/z$ and a bounded integrable function), and h_i similarly for g. For $i \leq r$ we have $h_i^{(n)} \longrightarrow h_i$ uniformly since

$$|h_i^{(n)}(\zeta) - h_i(\zeta)| \leq \frac{16}{\pi\delta}\left(\iint_U \frac{1}{|z-\zeta|^{3/2}}\,dx\,dy\right)^{2/3}\left(\iint_U |g_n(z) - g(z)|^3\,dx\,dy\right)^{1/3}$$

and the first integral is finite while the second tends to zero by dominated convergence.

Now $\|h_i\| \leq A_1\|g\| < A_1\epsilon$ so that if $n = n_\delta$ is chosen large enough $\|h_i^{(n)}\| < A_1\epsilon$, $i \leq r$. Evidently $h_i^{(n)}$ is analytic outside D_i and vanishes at ∞, so that $|h_i^{(n)}(\zeta)| < A_1\epsilon \min(1, \delta/|\zeta - z_i|)$, where z_i is the center of D_i by Schwarz's lemma. We fix $n = n_\delta$ (for our δ) and note that $g_n\phi_i + h_i^{(n)} \in A(U)$: indeed, if we extend g_n continuously to all the Riemann sphere S, then Green's formula shows

$$(4.3)\qquad g_n(\zeta)\phi_i(\zeta) + \frac{1}{\pi}\iint \frac{\partial\phi_i}{\partial\bar z}\frac{g_n(z)}{z-\zeta}\,dx\,dy = \frac{1}{\pi}\iint \frac{g_n(z)-g_n(\zeta)}{z-\zeta}\frac{\partial\phi_i}{\partial\bar z}\,dx\,dy,$$

and since the right side is evidently analytic where g_n is, in particular on U, and both sides are continuous, we only have to see

$$(4.4)\qquad \zeta \longrightarrow \frac{1}{\pi}\iint_{U'} \frac{\partial\phi_i}{\partial\bar z}\frac{g_n(z)}{z-\zeta}\,dx\,dy$$

lies in $A(U)$ (for our extended g_n) to see $g_n\phi_i + h_i^{(n)} \in A(U)$ from (4.2) and (4.3). But (4.4) defines a function analytic on U, and again continuous as a convolution of the locally integrable function $1/z$ and a bounded integrable function, so (4.4) lies in $A(U)$, as does $g_n\phi_i + h_i^{(n)}$.

We set $H_\delta = \Sigma_{i=1}^r (g_n\phi_i + h_i^{(n)})^3$ and can now assert that

$$(4.5)\qquad |H_\delta(\zeta)| \leq A_2 \min\left(1, \frac{\delta}{d(\zeta, F)}\right),\quad \zeta \in U^-,$$

and

(4.6) If $\zeta \in F$, $|1 - H_\delta(\zeta)| < A_3 < 1$ (for $\epsilon < A_4$).

For (4.5), write

$$H_\delta = g_n^3 \sum_{i=1}^r \phi_i^3 + 3g_n \sum_{i=1}^r \phi_i h_i^{(n)}(g_n\phi_i + h_i^{(n)}) + \sum_{i=1}^r (h_i^{(n)})^3$$
$$= T_1 + T_2 + T_3.$$

Now T_1 and T_2 involve at most 25 non-zero terms at any ζ which all vanish when $d(\zeta, F) > 2\delta$ (since each D_i meets F), so that these can be accommodated in (4.5) by simply choosing A_2 large. For T_3, since $|h_i^{(n)}| < A_2\epsilon \min(1, \delta/|\zeta - z_i|)$ we have

$$(4.7)\qquad |T_3| < A_2^3\epsilon^3 \sum_{i=1}^r \min\left(1, \frac{\delta^3}{|\zeta - z_i|^3}\right).$$

We estimate this by noting that (for ζ fixed) each D_i meets at least one (and at most two) of the circles $C_n = \{z : |z - \zeta| = n\delta\}$, $n = 1, 2, \cdots$, and if $N(n)$ denotes the number of D_i meeting C_n, and I is the corresponding index set in $\{1, 2, \cdots, r\}$ then

$$\sum_{i\in I} \chi_{D_i} \leqslant 25\chi_{B_n}$$

for the annulus $B_n = \{z\colon (n-2)\delta \leqslant |z-\zeta| \leqslant (n+2)\delta\}$ since no point lies in more than 25 of our discs D_i. Thus

$$\pi\delta^2 N(n) \leqslant 25((n+2)^2-(n-2)^2)\delta^2\pi, \quad \text{or} \quad N(n) \leqslant 200n,$$

and from (4.7), if j is the largest integer with $(j+1)\delta < d(\zeta, F)$, *and* $j \geqslant 2$ we have

$$|T_3(\zeta)| \leqslant A_2^3\,\epsilon^3 \sum_{n\geqslant j} 200n \cdot \min\left(1, \frac{\delta^3}{(n-1)^3\,\delta^3}\right)$$

since $C_1, C_2, \cdots, C_{j-1}$ miss each D_i and $|z_i - \zeta| \geqslant (n-1)\delta$ if $D_i \cap C_n \neq \varnothing$. So, if $j \geqslant 2$, we have

$$(4.8) \qquad |T_3(\zeta)| \leqslant A_2^3\,\epsilon^3 \sum_{n\geqslant j} \frac{200n}{(n-1)^3} < A_5\epsilon^3 \frac{1}{j+2} \leqslant A_5\epsilon^3 \frac{\delta}{d(\zeta, F)}$$

since $(j+2)\delta \geqslant d(\zeta, F)$. And if $j = 1$ then

$$(4.9) \qquad |T_3(\zeta)| \leqslant A_2^3\epsilon^3 \left(N(1) + \sum_{n=2}^{\infty} 200n \frac{\delta^3}{(n-1)^3\delta^3}\right)$$

so that we need only increase our constant to obtain (4.5).

For (4.6), we have $\|T_2\| \leqslant 75A_1\epsilon(1 + A_1\epsilon)$ since at most 25 ϕ_i do not vanish at any given point while $|g_n| \leqslant 1$ and $|h_i^{(n)}| < A_1\epsilon$. Since $|T_3| \leqslant A_6\epsilon$ from (4.8) and (4.9), $\|T_2 + T_3\| \leqslant A_7\epsilon$. Now for $\psi = \Sigma_{i=1}^r \phi_i^3$, since $\Sigma\phi_i = 1$ on F, if $\zeta \in F$ we have $\phi_i(\zeta) \geqslant (25)^{-1}$ for some i, so $\psi(\zeta) \geqslant (25)^{-3}$, $\zeta \in F$, and

$$\begin{aligned}|1 - H_\delta(\zeta)| &\leqslant A_7\epsilon + |1 - g_n(\zeta)^3\psi(\zeta)| \\ &\leqslant A_7\epsilon + (1 - \psi(\zeta)) + \psi(\zeta)|\,1 - g_n(\zeta)^3| \\ &\leqslant A_7\epsilon + 1 - (25)^{-3} + \psi(\zeta) \cdot 3|1 - g_n(\zeta)| \\ &\leqslant A_7\epsilon + 1 - (25)^{-3} + 3\epsilon, \qquad \zeta \in F,\end{aligned}$$

since $|1 - g_n| < \epsilon$ on F and $\psi \leqslant 1$; so

$$(4.10) \qquad |1 - H_\delta(\zeta)| < 1 - \tfrac{1}{2}(25)^{-3} \quad \text{if} \quad (A_7 + 3)\epsilon < \tfrac{1}{2}(25)^{-3}$$

which is (4.6).

Now as $\delta \longrightarrow 0$, $H_\delta \longrightarrow 0$ boundedly on $U^-\backslash F$ by (4.5). So for each integer k we can choose $\delta = \delta_k > 0$ so that

$$\left|\int_{U^-\backslash F} 1 - (1 - H_{\delta_k})^k\, d\sigma\right| < 1/k,$$

and since $\sigma \perp A(U)$ this says

$$\left|\int_F 1 - (1 - H_{\delta_k})^k\, d\sigma\right| < 1/k,$$

so that we can now let $k \longrightarrow \infty$ and get $\sigma F = 0$ from (4.10), our contradiction.

The estimate of $|T_3(\zeta)|$ in Davie's proof is a standard part of the Vituskin approach (in which the "interference" mentioned earlier is controlled); there analytic capacity conditions are used to remove terms of order < 3 in a Laurent expansion (cf. [Gam]); the introduction of a sum of cubes here is a device due to Øksendal [Ø].

Davie's proof also yields a result for $R(K)$. Recall that λ_Q denotes area measure restricted to Q, the set of non-peak-points for $R(K)$.

THEOREM 4.2 (DAVIE). *Let* $f \in L^\infty(\lambda_Q)$ *be a* w^* *limit of a bounded sequence in* $R(K)$. *Then there is a sequence* $\{f_n\}$ *in* $R(K)$ *with* $\|f_n\| \leq \|f\|$ *which converges* w^* *to* f *in* $L^\infty(\lambda_Q)$.

The proof is essentially the same as that of Theorem 4.1 with $R(K)$ in place of $A(U)$, K in place of U, and w^* convergence in $L^\infty(\lambda_Q)$ in place of pointwise convergence (with $g_n \to g$ a.e. (λ_Q) in the conclusion of Step I obtained by passing to a sequence of finite convex combinations), up to the definition of h_i^n. Here one defines

$$h_i^{(n)}(\zeta) = \frac{1}{\pi} \iint_Q \frac{\partial \phi_i}{\partial \bar{z}} \frac{g_n(z)}{z - \zeta}\, dx\, dy$$

(with h_i defined similarly for g) and $h_i^{(n)} \to h_i$ uniformly again because of Hölder and dominated convergence, and the problem here is to see $g_n \phi_i + h_i^{(n)} \in R(K)$. Now we can take each g_n actually rational (replacing it by a uniformly approximating rational function), and if we alter g_n off a neighborhood of K so as to be in $C(S)$ then

$$\begin{aligned} G(\zeta) &= \frac{1}{\pi} \iint_{\mathbf{C}} \frac{g_n(z) - g_n(\zeta)}{z - \zeta} \frac{\partial \phi_i}{\partial \bar{z}}\, dx\, dy \\ &= g_n(\zeta)\phi_i(\zeta) + \frac{1}{\pi} \iint_{\mathbf{C}} \frac{\partial \phi_i}{\partial \bar{z}} \frac{g_n(z)}{z - \zeta}\, dx\, dy \end{aligned}$$

is necessarily rational, so we only need to see

$$\zeta \to \iint_{Q'} \frac{\partial \phi_i}{\partial \bar{z}} \frac{g_n(z)}{z - \zeta}\, dx\, dy$$

is in $R(K)$. Of course the corresponding integral over K' is in $R(K)$, so we only have to see the continuous function

$$F(\zeta) = \iint_P \frac{\partial \phi_i}{\partial \bar{z}} \frac{g_n(z)}{z - \zeta}\, dx\, dy$$

is in $R(K)$, where P is the set of peak points for $R(K)$. But if the measure μ on K is orthogonal to $R(K)$ then

$$\begin{aligned} \int_K F(\zeta)\mu(d\zeta) &= \iint_P \left(\int_K \frac{\partial \phi_i}{\partial \bar{z}} \frac{g_n(z)}{z - \zeta} \mu(d\zeta) \right) dx\, dy \\ &= \iint_P \frac{\partial \phi_i}{\partial \bar{z}} g_n(z) \left(\int_K \frac{1}{z - \zeta} \mu(d\zeta) \right) dx\, dy \end{aligned} \tag{4.11}$$

(where Fubini applied since $1/z$ is locally integrable). Now it is known that $\int (1/|z - \zeta|)\, |\mu|(d\zeta) < \infty$ a.e. $(dx\, dy)$ and (Wilken's theorem) that if that happens and

$\int(1/(z-\zeta))\mu(d\zeta) \neq 0$ then z is not a peak point for $R(K)$ [Br]. So the inner integral of (4.11) vanishes a.e. and $\mu \perp F$, whence $F \in R(K)$.

The remainder of the proof is unchanged.

Actually 4.2 applies to every f in $H^\infty(\lambda_Q)$ [Gam–Gar]:

COROLLARY 4.3. *If* $f \in H^\infty(\lambda_Q)$ *there is a sequence* $\{f_n\}$ *in* $R(K)$ *with* $\|f_n\| \leqslant \|f\|$, *which converges* w^* *to* f.

Indeed, if B_R denotes the subspace of $H^\infty(\lambda_Q)$ of those elements f satisfying the hypotheses of 4.1 then ball B_R is w^* closed since the w^* topology of ball $H^\infty(\lambda_Q)$ is metrizable (with metric ρ say), and, for any sequence $\{f_n\}$ in ball $B_R, f_n \longrightarrow f$ w^*, we can find elements g_n of ball $R(K)$ with $\rho(f_n, g_n) < 1/n$ by 4.2, so that $g_n \longrightarrow f$ w^* and $f \in B_R$. Thus by Kreĭn-Smulian B_R is w^* closed; it contains $R(K)$, and so coincides with $H^\infty(\lambda_Q)$, the w^* closure of $R(K)$.

Actually, w^* convergence of a bounded sequence in $R(K)$ to f in $L^\infty(\lambda_Q)$ implies the sequence converges pointwise on Q to a function $g = f$ a.e. λ_Q with g independent of the particular sequence, as Davie [D] shows. In [Gam–Gar] Gamelin and Garnett use the last version of Davie's result to show not only that $R(K) = C(K) \cap L^\infty(\lambda_Q)$ (in both a constructive and another, abstract, way) but also the more general result that one has, for $h \in C(K)$,

$$\|h - R(K)\|_\infty = \|h - H^\infty(\lambda_Q)\|_\infty,$$

extending a result of Sarason.

Cole (unpublished) observed that a result like Davie's could be used to show any $z \in Q$ had a representing measure $\ll \lambda_Q$, and Gamelin and Garnett show that in fact such measures are w^* dense in M_z, here the representing measures for $R(K)$ carried by K. One of the measures on ∂K which represents $z \in K$ is the harmonic measure λ_z, and it can easily occur that when $K = U^-$, U open, ∂K has positive area; interestingly, in [Ø] Øksendal shows λ_z is always singular with respect to the area measure.

For one interested in further results on rational approximation, the papers of Davie, Gamelin, Garnett, and Øksendal in the bibliography will provide a good guide to the current literature.

5. CHARACTERIZATIONS OF $C(X)$

What properties of $C(X)$ can its separating proper subalgebras A share? Clearly A cannot provide too many peak sets; on the other hand it can provide the same peak points, even with X the spectrum of A, or the same multiplicative measures, as Cole's example shows. There are a variety of properties which they cannot share, and which therefore characterize $C(X)$; recent among these are various results of Bernard which extend earlier results of Hoffman and Wermer [H–W] and Wermer [W], which we will now discuss. Bernard's results have appeared in scattered C. R. notes, and are covered together in a paper soon to appear [Be 5]; the proofs below are precisely Bernard's.

We begin with Bernard's simple proof of a theorem of Hoffman and Wermer.

THEOREM 5.1. *Suppose* $1 \in A$, *a closed separating subalgebra of* $C(X)$. *Then* $\mathrm{Re}\, A$ *is uniformly closed only if* $A = C(X)$.

DEFINITION. *By a Banach subalgebra* A *of* $C(X)$ *we shall mean a subalgebra which is a Banach algebra under some norm* N.

Of course then $|a(x)| \leq N(a)$ since multiplicative linear functionals are of norm ≤ 1, and so $\|a\| \leq N(a)$. (We will reserve $\| \ \|$ for the sup norm here.) At one point it will be useful to know that we really only need our algebra A to be a Banach space under a norm N which makes the injection of A into $C(X)$ continuous: in fact then $a \longrightarrow ab$ is N-continuous by the closed graph theorem so $N(ab) \leq c_b N(a)$, and category then shows $N(ab) \leq CN(a)$ for $N(b) \leq 1$, whence $N(ab) \leq CN(a)N(b)$, and we have an equivalent Banach algebra norm.

LEMMA 5.2. *Suppose* $1 \in A$, *a Banach subalgebra of* $C(X)$ *which separates* X. *Then if* $\mathrm{Re}\, A$ *is uniformly closed,* $A_R = A \cap C_R(X)$ *separates* X.

By the open mapping theorem we have a k for which

$$(5.1) \qquad u \in \mathrm{Re}\, A,\ \|u\| \leq 1 \Rightarrow u + iv \in A,\ v \in C_R(X),\ N(u + iv) \leq k.$$

Now let K be a set of constancy of the real algebra A_R; we assert that *if* K *is not a singleton then there are* x_1, x_2 *in* K *and* $f \in A$ *with*

$$(5.2) \qquad f(x_1) = 0,\ f(x_2) = (2k + 1)i, \quad |\mathrm{Re}\, f| \leq 1.$$

Indeed choose $g \in A$ and $x_1, x_2 \in K$ with $g(x_1) = 0, 1 = g(x_2) = \|g\|_K$ ($= \sup |g(K)|$). For $\epsilon > 0$ let $H = \{x : |g(x)| \geq 1 + \epsilon\}$ so $H \cap K = \varnothing$. Since K is a peak set for A_R^- ($= C(Y)$ for some quotient Y of X by Stone-Weierstrass), we can choose a high power of the peaking function as a $g' \in A_R^-$ with $\|g'g\| < 1 + \epsilon, g' \equiv 1$ on K. For an approximating g_0 in A_R with $g_0(x_2) = 1 = g'(x_2)$ we will also have $\|g_0 g\| < 1 + \epsilon$, and so for $h = g_0 g \in A$, $h(x_1) = 0$, $h(x_2) = 1$, $\|h\| < 1 + \epsilon$.

Now with $\epsilon > 0$ sufficiently small we can find a polynomial p which maps the disc $(1 + \epsilon)D$ into $|\text{Re } z| < 1$ and has $p(0) = 0, p(1) = (2k + 1)i$. (We can choose a conformal map ϕ of D° onto $|\text{Re } z| < 1$ which is 0 at 0, imaginary on $R \cap D$, and has $\text{Im } \phi(z) \longrightarrow + \infty$ for $z \longrightarrow 1$ so that for ϵ small $z \longrightarrow \phi(z/(1 + 2\epsilon))$ has a large imaginary value at $z = 1$, approximate this on $(1 + \epsilon)D$ by a polynomial still 0 at 0 and imaginary on $R \cap D$, and take p as a real multiple.) Thus $f = p \circ h \in A$ will yield our assertion.

Now if A_R does not separate X we have such a K, x_1, x_2 and f. By (5.1) for $u = \text{Re } f$ we have a $v \in C_R(X)$ with $u + iv \in A$, $N(u + iv) \leqslant k$, so $\|u + iv\| \leqslant k$, and $a = (1/i)\,(f - (u + iv)) \in A_R$ while

$$|a(x_1)| = |v(x_1)| \leqslant \|v\| \leqslant \|u + iv\| \leqslant k$$

and

$$|a(x_2)| \geqslant 2k + 1 - |v(x_2)| \geqslant k + 1$$

so that $a(x_1) \neq a(x_2)$ despite the fact that $x_1, x_2 \in K$. Our proof of 5.2 is complete.

The proof of the Hoffman-Wermer Theorem 5.1 is now immediate: Since A_R is dense in $C_R(X)$ by Stone-Weierstrass, and closed since $A_R = A \cap C_R(X)$ and A is, $A_R = C_R(X)$. But of course that says $A \supset A_R + iA_R = C(X)$.

As a corollary to the Hoffman-Wermer theorem Bernard obtained a far reaching generalization.

THEOREM 5.3. *Suppose* $1 \in A$, *a Banach subalgebra of* $C(X)$ *which separates* X. *Then* $\text{Re } A$ *is uniformly closed only if* $A = C(X)$.

This extends (and no doubt was suggested by) an earlier result of Sidney and Stout [S–S], that for a separating uniform algebra A, $\text{Re } A\,|\,F$ is closed in $C(F)$ only if $A\,|\,F = C(F)$ (take $A\,|\,F$ as the Banach subalgebra of $C(F)$, normed of course as the isomorphic quotient algebra A/kF). The fundamental step in 5.3 is the remarkably simple and powerful

LEMMA 5.4 (BERNARD'S LEMMA). *Suppose* $E \subset F$, *each normed spaces, with the injection continuous. Let* $\tilde{E} = l_\infty(N, E)$ (*the bounded functions from* $N = \{1, 2, \cdots\}$ *to* E, *normed as usual*), $\tilde{F} = l_\infty(N, F)$. *Then* $\tilde{E} \subset \tilde{F}$, *and if* E *is a Banach space,* $\tilde{E}$ *dense in* $\tilde{F}$ *implies* $E = F$.

That $\tilde{E} \subset \tilde{F}$ is trivial since $E \overset{\subset}{\longrightarrow} F$ is continuous. From the final portion of the usual proof of the open mapping theorem one can observe that to see $E = F$ it suffices to see that for some n, nB_E, the n-ball in E, is ½-dense in B_F (i. e., for $y \in B_F$ there is an $x \in nB_E$ with $\|x - y\|_F < ½$). And if that fails, for each n we have a y_n in B_F with $\|y_n - nB_E\|_F \geqslant ½$. But by hypothesis there is an $x = \{x_n\}$ in $\tilde{E}$ with $\|x_n - y_n\| < ¼$ for all n, and so $x_n \notin nB_E$ and $\|x_n\|_E > n$ for each n, so that $x \notin \tilde{E}$.

Bernard's proof of Theorem 5.3 is now as follows: since $\text{Re } A$ is closed, $\text{Re } A^- \subset \text{Re } A$, where A^- is the uniform closure of A, so $\text{Re } A^- = \text{Re } A$ is closed. Thus the

Hoffman-Wermer theorem says $A^- = C(X)$, whence $\operatorname{Re} A^- = \operatorname{Re} A = C_R(X)$. So by the open mapping theorem we have a k for which

(5.3) $\qquad u \in C_R(X)$ implies $u + iv \in A$, $v \in C_R(X)$, $N(u + iv) \leq k\|u\|$.

Now $\widetilde{A} = l_\infty(N, A)$ is a subalgebra of $l_\infty(N, C(X))$, and we can identify the latter with $C(\beta(N \times X))$, so (5.3) allows us to assert that

(5.4) $$\operatorname{Re} \widetilde{A} = C_R(\beta(N \times X)):$$

for with $u = \{u_n\} \in \operatorname{Re} l_\infty(N, C(X)) = l_\infty(N, C_R(X))$ we have $v_n \in C_R(X)$, $N(u_n + iv_n) \leq k\|u\|_\infty = k \sup \|u_n\|$, $u_n + iv_n \in A$, so that $\widetilde{a} = \{u_n + iv_n\} \in \widetilde{A}$ and $\operatorname{Re} \widetilde{a} = u$.

Since (5.4) clearly implies $\widetilde{A}$ separates $\beta(N \times X)$, Hoffman-Wermer applies to $\widetilde{A}^-$ (since $\operatorname{Re} \widetilde{A}^- = \operatorname{Re} \widetilde{A} = C_R(\beta(N \times X)))$. Thus $\widetilde{A}^- = C(\beta(N \times X)) = l_\infty(N, C(X))$ and that says $A = C(X)$ by Bernard's lemma.

One application of interest occurs in harmonic analysis. If Γ is a compact abelian group, the dual of a discrete group G, a closed subset H of Γ is called a Helson set if $l_1(G)^\wedge \mid H = C(H)$. Now if S is a subsemigroup of G with $S \cup (-S) = G$ then $l_1(S)^\wedge | H = C(H)$ for any Helson set H: for $\operatorname{Re}(l_1(S)^\wedge | H) = \operatorname{Re}(l_1(G)^\wedge | H) = \operatorname{Re} C(H) = C_R(H)$, as is easily seen, so $l_1(S)^\wedge | H = C(H)$ by Theorem 5.3. (A special case is known as Wik's theorem.)

Bernard has exploited the kind of argument used to prove Theorem 5.3, as well as the result itself, to obtain a variety of consequences, in particular an extension of Wermer's theorem that if A is uniformly closed and separating $\operatorname{Re} A$ cannot be an algebra unless $A = C(X)$, and, recently, that $\operatorname{Re} A$ cannot be a lattice. In this the following notion is useful.

DEFINITION. *$A \subset C(X)$ is called ultraseparating if $\widetilde{A}$ separates $\beta(N \times X)$ (and ultraseparating on $K \subset X$ if $\widetilde{(A|K)}$ separates $\beta(N \times K)$).*

In particular both occur if a uniform algebra A "approximates in modulus", i.e., given $\epsilon > 0$ and $f > 0$ in $C_R(X)$ there is an $a \in A$ with $|f - |a|| < \epsilon$: for if $f \in C_R(\beta(N \times X))$ is zero at one point x^* of $\beta(N \times X)$ and 1 at another, y^*, $0 \leq f \leq 1$, then as an element of $l_\infty(N, C_R(X))$ $f = \{f_n\}$, and we can find a_n in A with $|f_n - |a_n|| < 1/3$ so that $\widetilde{a} = \{a_n\} \in \widetilde{A}$, and $\widetilde{a}$ (extended to $\beta(N \times X)$) has a value of modulus $\geq 2/3$ at y^*, $\leq 1/3$ at x^*. Finally, it is easy to see ultraseparation implies ultraseparation on closed subsets.

LEMMA 5.5. *If $1 \in A$, a Banach subalgebra of $C(X)$, A is ultraseparating and $B \supset A$ is another Banach subalgebra of $C(X)$ which is conjugate closed, then $B = C(X)$.*

Since the point evaluations $a \longrightarrow a(x)$ and $b \longrightarrow b(x)$ are continuous on A and B the closed graph theorem shows the inclusion $A \xrightarrow{\subset} B$ is continuous. As a consequence $\widetilde{A} \subset \widetilde{B} (\subset l_\infty(N, C(X)))$ and thus $\widetilde{B}$ separates $\beta(N \times X)$. Moreover the closed graph theorem shows conjugation is continuous on B, and that implies $\widetilde{B}$ is conjugate closed (in $l_\infty(N, C(X))$, hence in $C(\beta(N \times X)))$. So $\widetilde{B}$ is dense in $C(\beta(N \times X))$ by Stone-Weierstrass, or $\widetilde{B}$ is dense in $\widetilde{C(X)}$, whence $B = C(X)$ by Bernard's Lemma 5.4.

THEOREM 5.6. *If* $1 \in A$, *and* A *is an ultraseparating Banach subalgebra of* $C(X)$ *and* $\operatorname{Re} A$ *is closed under multiplication then* $A = C(X)$.

We have only to take $B = \operatorname{Re} A + i \operatorname{Re} A$, which is by hypothesis a subalgebra of $C(X)$ containing A, and also a Banach space which injects continuously into $C(X)$ when normed by $N(\operatorname{Re} a + i \operatorname{Re} b) = N_1(\operatorname{Re} a) + N_1(\operatorname{Re} b)$, where $N_1(\operatorname{Re} a) = \|a + E\|$ is the quotient norm in the real Banach space A/E and E is the kernel of $a \longrightarrow \operatorname{Re} a$. By our preliminary remarks, B is a Banach subalgebra of $C(X)$ and from Lemma 5.5 we have $B = \operatorname{Re} A + i \operatorname{Re} A = C(X)$, so $\operatorname{Re} B = \operatorname{Re} A = C_R(X)$. And that implies $A = C(X)$ by Bernard's Theorem 5.3.

COROLLARY 5.7 (WERMER). *If* A *is a closed separating subalgebra of* $C(X)$ *containing the constants and* $\operatorname{Re} A$ *is an algebra then* $A = C(X)$.

For $\operatorname{Re} A$ is necessarily dense in $C_R(X)$, so that A "approximates in modulus".

COROLLARY 5.8. *If* $1 \in A$, *a closed subalgebra of* $C(X)$ *which approximates in modulus, then* $\operatorname{Re}(A|K)$ *is closed under multiplication only if* $A|K = C(K)$.

For $A|K$ is ultraseparating on K as noted earlier. For the disc algebra $A = A(D)$ this has the following consequence: if $K \subset T$ is compact and has positive measure then there is a real continuous function u on K which has an extension in $\operatorname{Re} A(D)$ while u^2 does not. (Alternatively, u has a continuous extension to D harmonic on D° which has a continuous harmonic conjugate, while u^2 does not.)

Most recently Bernard has shown $\operatorname{Re} A$ cannot be a lattice when A is a proper closed separating subalgebra of $C(X)$. This depends on the following simple result of de Leeuw and Katznelson.

THEOREM 5.9. *Suppose* E *is a closed separating subspace of* $C_R(X)$ *and* $1 \in E$. *If a non-affine continuous* $\phi : R \longrightarrow R$ *operates on* E (*i. e. if* $\phi \circ E \subset E$) *then* $E = C_R(X)$.

PROOF. Trivially the convolution $k * \phi = \phi'$ will operate as well since ϕ' is a uniform limit on compacta of linear combinations of translates of ϕ (all of which operate since $1 \in E$!) and E is uniformly closed. Taking k to be C^∞ with small support we can insure that ϕ' is close enough to ϕ to also be non-affine, and C^∞. Replacing ϕ' by $\phi_1(t) = a\phi(t - t_0) + bt + c$ for appropriate a, b, c, t_0 we can assume $\phi_1(0) = \phi_1'(0) = 0$, $\phi_1''(0) = 2$, so $\phi_1(t) = t^2 + t^2\epsilon(t)$, where $\epsilon(t) \longrightarrow 0$ as $t \longrightarrow 0$.

But for any integer n,

$$\phi_n(t) = n^2 \phi_1(t/n) = t^2 + t^2 \epsilon(t/n)$$

operates; so finally t^2 operates, i.e., E is a closed separating subalgebra of $C_R(X)$, $1 \in E$, so $E = C_R(X)$ by Stone-Weierstrass.

Suppose now $E \subset C_R(X)$ is normed so the injection into $C_R(X)$ is continuous. Then $\phi : R \longrightarrow R$ *operates boundedly on* E if ϕ is continuous and for each $\alpha > 0$ there is an M for which

(5.5) $$N(u) < \alpha \text{ implies } \phi \circ u \in E \text{ and } N(\phi \circ u) \leqslant M.$$

THEOREM 5.10. *If* A *is a closed proper separating subalgebra of* $C(X)$ *and* $1 \in A$ *only affine functions operate boundedly on* $\operatorname{Re} A$ *(taken with its natural quotient norm).*

THEOREM 5.11. *If* A *is a closed proper separating subalgebra of* $C(X)$ *and* $1 \in A$, $\operatorname{Re} A$ *is not a lattice.*

Theorem 5.11 will be obtained from the argument leading to Theorem 5.10 by showing that if $\phi(t) = |t|$ operates on $\operatorname{Re} A$ it must operate boundedly. We begin with the main connecting link to the earlier results.

LEMMA 5.12. *Suppose* E *is a (real) Banach space which injects continuously into* $C_R(X)$ *as above, contains* 1 *and is ultraseparating on* X. *If* $E \neq C_R(X)$ *then only affine functions operate boundedly on* E.

Trivially if ϕ operates on E then ϕ operates on E^- in $C_R(X)$: if $u_n \in E$ and $u_n \longrightarrow f$ in $C_R(X)$ then $\phi \circ u_n \in E$ and $\phi \circ u_n \longrightarrow \phi \circ f$ uniformly, so $\phi \circ f \in E^-$. Since ϕ operates boundedly on E, it operates on $\widetilde{E} = l_\infty(N, E) \subset \widetilde{C_R(X)} = C_R(\beta(N \times X))$, hence it operates on the closure F of $\widetilde{E}$ in $C_R(\beta(N \times X))$. But ϕ non-affine implies $F = C_R(\beta(N \times X))$ by Katznelson and de Leeuw's Theorem 5.9, which applies since $\widetilde{E}$ separates $\beta(N \times X)$, i.e. E is ultraseparating. Hence $\widetilde{E}$ is dense in $\widetilde{C_R(X)}$, and $E = C_R(X)$ by Bernard's lemma.

In case $\operatorname{Re} A$ is dense in $C_R(X)$ the lemma applies to $E = \operatorname{Re} A$ since then $\operatorname{Re} A$ is ultraseparating. To prove 5.10, suppose ϕ is non-affine and operates boundedly on $\operatorname{Re} A$. Then it operates on $(\operatorname{Re} A)^-$, so since $(\operatorname{Re} A)^-$ separates, Theorem 5.9 implies $(\operatorname{Re} A)^- = C_R(X)$. Now $\operatorname{Re} A$ is ultraseparating on X so Lemma 5.12 applies to show $\operatorname{Re} A = C_R(X)$. By Bernard's theorem, $A = C(X)$.

To prove Theorem 5.11 we need another lemma.

LEMMA 5.13. *Suppose* $E \subset C_R(X)$ *is a Banach space which injects continuously into* $C_R(X)$, *contains* 1, *separates* X, *and is a lattice. Then there is a finite subset* F *of* X *such that* $x \notin F$ *implies* x *has a compact neighborhood* K_x *for which* $\phi(t) = |t|$ *operates boundedly on* $E|K_x$ *(taken with the natural quotient norm).*

For a compact $K \subset U$ open, an easy compactness argument (using the fact that E is a lattice which separates, and $0, 1 \in E$) shows that any element u of $E|K$ has an extension $\widetilde{u}$ in $E \equiv 0$ off U. (For $0 \leqslant u \leqslant 1$ it suffices to find a v in E, $= 1$ on K and $\equiv 0$ off U, and take $\widetilde{u} = u' \wedge v$, where $u'|K = u$.) By the open mapping theorem then there is a constant $k_{U,K}$ for which, for each $u \in E|K$, there is a $\widetilde{u}$ as above with

$$N(\widetilde{u}) \leqslant k_{U,K} N_K(u) \tag{5.6}$$

where $N_K(u) = \inf \{N(u') : u' = u \text{ on } K,\ u' \in E\}$.

Suppose the lemma fails and let E be the infinite set of x for which ϕ fails to operate boundedly on $E|K_x$ for each compact neighborhood K_x of x. Let x_0 be a cluster point of E. For $x_1 \in E \backslash \{x_0\}$ let K_1 be a compact neighborhood of x_1 in $E \backslash \{x_0\}$, and choose x_2 and a compact neighborhood K_2 of x_2 in $E \backslash (K_1 \cup \{x_0\})$. Continuing we obtain a sequence of disjoint compact neighborhoods K_n for which

$\phi(t) = |t|$ does not operate boundedly on $E|K_n$; we now choose open $U_n \supset K_n$ so that the U_n are also pairwise disjoint.

For each n we have $u_{n,k} \in E|K_n$ with $N_{K_n}(u_{n,k}) \longrightarrow 0$ as $k \longrightarrow \infty$ while all extensions of $|u_{n,k}|$ in E have norm $\geqslant n$, and thus in view of (5.6) we have elements $\widetilde{u}_n$ of E with $\widetilde{u}_n \equiv 0$ off U_n, $N(\widetilde{u}_n) < 2^{-n}$, while each extension of $|\widetilde{u}_n|\,|K_n$ in E has norm $\geqslant n$. But consider $\widetilde{u} = \Sigma \widetilde{u}_n$. It lies in E, so $|\widetilde{u}| \in E$, while $|\widetilde{u}|\,|K_n = |\widetilde{u}_n|\,|K_n$, so that $N(|\widetilde{u}|) \geqslant n$ for all n, the desired contradiction.

We can now prove 5.11. Suppose $\operatorname{Re} A$ is a lattice, and thus dense in $C_R(X)$. Then $\operatorname{Re} A|K$ is dense in $C_R(K)$, so $\operatorname{Re} A|K$ is ultraseparating on K. Applying 5.13 to $E = \operatorname{Re} A$ we have a finite $F \subset X$ with $\phi(t) = |t|$ operating boundedly on $\operatorname{Re} A|K_x, x \notin F$, where K_x is a neighborhood of x. And since Theorem 5.10 applies to $\operatorname{Re} A|K_x$ to show $\operatorname{Re} A|K_x = C_R(K_x)$, we know $A|K_x = C(K_x)$ by Bernard's theorem, for each $x \notin F$. It is now easy to see $A = C(X)$. (Consider the essential set ! [Br, p. 144]).

For further results along these lines the reader should see [Be 5].

BIBLIOGRAPHY

At the time of the conference two texts on function algebras were available:

[Br] A. Browder, *Introduction to function algebras,* Benjamin, New York, 1969. MR **39** #7431.

[Gam] T. W. Gamelin, *Uniform algebras,* Prentice-Hall, Englewood Cliffs, N. J., 1969.

Of these Browder's is the more elementary, Gamelin's the more succinct and extensive. A beginner might well approach them in sequence. Both are concerned with rational approximation, but only Gamelin covers Vituskin's work (with some minor flaws in the presentation; one can also consult Zalcman's monograph). Two more texts have since appeared:

J. Wermer, *Banach algebras and several complex variables,* Markham, Chicago, Ill., 1971.

which approaches approximation problems for subsets of $\mathbf{C}^n$, and includes results on analytic structure in spectra, and

E. L. Stout, *The theory of uniform algebras,* Bogden and Quigley, Tarrytown on Hudson, New York, 1971.

The following are referred to explicitly above

[Ba] R. F. Basener, *An example concerning peak points,* Brown University, Providence, R. I. (preprint).

[Be 1] A. Bernard, *Une caractérisation de* $C(X)$ *parmi les algèbres de Banach,* C. R. Acad. Sci. Paris Sér. A-B **267** (1968), A634–A635. MR **38** #2601.

[D] A. M. Davie, *Bounded limits of analytic functions,* University of California, Los Angeles (preprint).

[Du] A. Dufresnoy, *Parties réelles de certains quotients d'algèbres uniform,* C. R. Acad. Sci. Paris (to appear).

[Gam–Gar] T. W. Gamelin and J. Garnett, *Bounded approximation by rational functions,* University of California, Los Angeles (preprint).

[Be 2] A. Bernard, *Comparison d'algèbres de fonctions à l'aide des parties réelles de leurs éléments,* C. R. Acad. Sci. Paris Sér. A-B **270** (1970), A29–A32. MR **41** #4245.

[Be 3] ———, *Algèbres ultraséparantes de fonctions continues,* C. R. Acad. Sci. Paris Sér. A-B **270** (1970), A818–A819. MR **41** #826.

[Be 4] ———, *Fonctions qui opèrent sur* Re A, C. R. Acad. Sci. Paris Sér. A-B **271** (1970), A1120–A1121. MR **42** #5048.

[Be 5] ———, *Espace des parties réelles des èléments d'une algèbre de Banach de fonctions,* J. Functional Analysis (to appear).

[D–S] N. Dunford and J. T. Schwartz, *Linear operators.* I: *General theory,* Pure and Appl. Math., vol. 7, Interscience, New York, 1958. MR **22** #8302.

[Gar–Gl] J. Garnett and I. Glicksberg, *Algebras with the same multiplicative measures,* J. Functional Analysis **1** (1967), 331–341. MR **36** #691.

[Gl] I. Glicksberg, *The abstract F. and M. Riesz theorem,* J. Functional Analysis **1** (1967), 109–122. MR **35** #2146.

[H] K. Hoffman, *Banach spaces of analytic functions,* Prentice-Hall Series in Modern Analysis, Prentice-Hall, Englewood Cliffs, N. J., 1962. MR **24** #A2844.

[H–W] K. Hoffman and J. Wermer, *A characterization of* $C(X)$, Pacific J. Math. **12** (1962), 941–944. MR **27** #325.

[K–S] H. König and G. Seever, *The abstract F. and M. Riesz theorem,* Duke Math. J. **36** (1969), 791–797.

[L 1] G. Lumer, *Algèbres de fonctions et espaces de Hardy,* Lecture Notes in Math., no. 75, Springer-Verlag, Berlin and New York, 1968. MR **39** #7434.

[L 2] ———, *Espaces de Hardy en pliesiers variable complexes,* C. R. Acad. Sci. Paris (to appear); *Hardy spaces in several complex variables* (in preparation).

[L 3] ———, *On Wermer's maximality theorem,* Invent. Math. **8** (1969), 236–237. MR **40** #4769.

[R] J. Rainwater, *A note on the preceding paper,* Duke Math. J. **36** (1969), 798–800.

[S–S] S. J. Sidney and E. L. Stout, *A note on interpolation,* Proc. Amer. Math. Soc. **19** (1968), 380–382. MR **36** #6944.

[V 1] N. Th. Varopoulos, *Ensembles pics et ensembles d'interpolation pour les algèbres uniformes,* C. R. Acad. Sci. Paris (to appear).

[V 2] ———, *Sur la reunion de deux ensembles d'interpolation d'une algèbre uniforme,* C. R. Acad. Sci. Paris (to appear).

[W] J. Wermer, *The space of real parts of a function algebra,* Pacific J. Math. **13** (1963), 1423–1426. MR **27** #6152.

[Z] L. Zalcman, *Analytic capacity and rational approximation,* Lecture Notes in Math., no. 50, Springer-Verlag, Berlin and New York, 1968. MR **37** #3018.

The following papers, while not specifically referred to (perhaps!) have appeared since the bibliography of [Gam] was formed. For earlier work the reader is referred to that bibliography.

P. Ahern, *A condition for peak points,* Duke Math. J. **37** (1970), 67–70. MR **41** #2025.

H. Alexander, *Polynomial approximation and analytic structure* (to appear).

C. Berger and W. Saffern, *Dirichlet enbalgebres of the disc algebra* (preprint).

J. -E. Björk, *Analytic structure in the maximal ideal space of a uniform algebra,* Ark. Mat. **8** (1970), 239–244;

A. Browder, *States, numerical range, etc.* Brown University, Providence, R. I. (preprint).

A. M. Davie, *Real annihilating measures for* $R(K)$, J. Functional Analysis **6** (1970), 357–386.

A. M. Davie, *Bounded approximation and dirichlet sets,* J. Functional Analysis **6** (1970), 460–467.

———, *Dirichlet algebras of analytic functions,* J. Functional Analysis **6** (1970), 348–356. MR **42** #2300.

———, *Analytic capacity and approximation problems,* University of California, Los Angeles (preprint).

———, *Linear extension operators for spaces and algebras of functions,* University of California, Los Angeles (preprint).

A. M. Davie, T. W. Gamelin and J. Garnett, *Distance estimates and pointwise bounded density,* University of California, Los Angeles (preprint).

A. M. Davie and A. Stray, *Interpolation sets for analytic functions,* University of California, Los Angeles (preprint).

A. M. Davie and B. K. Øksendal, *Rational approximation on the union of sets,* University of California, Los Angeles (preprint); Proc. Amer. Math. Soc. (to appear).

———, *Peak interpolation sets for some algebras of analytic functions,* University of California, Los Angeles (preprint).

S. Fisher, *Norm-compact sets of representing measures,* Proc. Amer. Math. Soc. **19** (1968), 1063–1068. MR **38** #1531.

T. W. Gamelin, *Uniform algebras containing the polynomials* (to appear).

———, *Norm compactness of representing measures for* $R(K)$, J. Functional Analysis **3** (1969), 495–500. MR **39** #782.

———, *Localization of the corona problem* (to appear).

T. W. Gamelin and J. Garnett, *Constructive techniques in rational approximation,* Trans. Amer. Math. Soc. **143** (1969), 187–200. MR **40** #2882.

———, *Pointwise bounded approximation and hypodirichlet algebras,* Bull. Amer. Math. Soc. **77** (1971), 137–141. MR **42** #6491.

———, *Distinguished homeomorphisms and fiber algebras,* Amer. J. Math. **92** (1970), 455–474.

T. W. Gamelin and D. R. Wilken, *Closed partitions of maximal ideal spaces,* Illinois J. Math. **13** (1969), 789–795. MR **40** #4767.

J. Garnett, *Metric conditions for rational approximation,* Duke Math. J. **37** (1970), 689–699. MR **42** #4746.

———, *Interpolating sequences for bounded harmonic functions,* University of California, Los Angeles (preprint).

H. König, *Über das von Neumannsche Minimax-Theorem,* Arch. Math. **19** (1968), 482–487. MR **39** #1947.

A. Stray, *An approximation theorem for subalgebras of* H^∞, Pacific J. Math. **35** (1970), 511–515.

———, *Approximation and interpolation,* Pacific J. Math. (to appear).

———, *Interpolation by analytic functions* (preprint).